D1527970

Satellite Broadcast Systems Engineering

For a complete listing of the *Artech House Space Technology and Applications Series*, turn to the back of this book.

Satellite Broadcast Systems Engineering

Jorge Matos Gómez

Artech House
Boston • London
www.artechhouse.com

Library of Congress Cataloging-in-Publication Data
Matos Gómez, Jorge.
    Satellite broadcast systems engineering/Jorge Matos Gómez.
        p.   cm. — (Artech House space technology and applications series)
    Includes bibliographical references and index.
    ISBN 1-58053-313-2 (alk. paper)
    1. Artificial satellites in telecommunication.   I. Title   II. Series.
TK5104 .M39 2002
621.382'5—dc21   2001056662

British Library Cataloguing in Publication Data
Matos Gómez, Jorge
    Satellite broadcast systems engineering.—(Artech House space technology and applications
    series)
    1. Television broadcasting   2. Artificial satellites in telecommunication
    I. Title
    621.3'8853

    ISBN 1-58053-313-2

Cover design by Yekaterina Ratner

© 2002 ARTECH HOUSE, INC.
685 Canton Street
Norwood, MA 02062

International Standard Book Number: 1-58053-313-2
Library of Congress Catalog Card Number: 2001056662

10 9 8 7 6 5 4 3 2 1

To the memory of my father and mother

Contents

Preface

This work is an outgrowth of my lecture notes on satellite broadcast systems for the senior undergraduate and graduate programs at Cuba's Facultad de Ingenieria Electrica (FIE) at the Universidad Central de Las Villas (UCLV) and other institutions abroad. The purpose of this book is to provide readers with the basic concepts and engineering tools needed for calculations in the exciting field of satellite broadcast systems.

The understanding of a complete satellite broadcast system requires coverage of a broad range of topics. It is assumed that readers are familiar with basic communication circuits and systems (including modulation, channel coding, and noise), antennas, and microwave propagation. Lengthy theoretical derivations are avoided, but mathematical and numerical results are used and explained to clarify basic concepts. Most of the calculations for satellite system performance are carried out in decibels or related units. These units and calculations are explained, but as a prerequisite, readers should have a basic working knowledge of the decibel.

This book aims to discuss the main elements of satellite television systems. It consists of seven chapters. Chapter 1 deals with the structure and basic aspects of the satellite broadcast system, ending with a discussion of WARC'77 and RARC'83 and their frequency-orbit planning aspects and other related parameters and recommendations. Chapter 2 provides an overview of geostationary satellites used in broadcast systems, emphasizing satellites that use nonregenerative transponders. Readers can properly use this chapter in association with satellite systems handbooks. Chapter 3 describes the basic engineering equations needed to analyze and design the radio transmission link to and from geostationary satellites. It also considers rain's influence on signal reception and provides an overview of other impairments. Chapter 4 deals with standards used today in satellite TV transmissions, both analog and digital. The key emphasis is on the digital standard for digital broadcast video (DBV-S). Because there are many books and references on the Motion Picture Expert Group (MPEG), only an overview of this compression technique is provided. Chapter 5 deals with TV reception system engineering, including its interference aspects. Chapter 5 also includes a general guideline for antenna size selection (diameter). Subsequently, Chapter 6 introduces DAB systems and outlines some projected systems. Worked examples are included in almost all chapters to illustrate key concepts and basic engineering methods.

Acknowledgments

Much of the material for this book was gathered while I was on sabbatical leaves at École Nationale Supérieure des Télécommunications (ENST) site in Toulouse, France, and at Decom-Faculdad de Engenharia Elétrica e de Computação (FEEC)-Universidade Estadual de Campinas (UNICAMP) in Brazil. My thanks to Professor Gerald Maral (ENST) and Professor Renato Baldini (Decom-FEEC-UNICAMP) for providing me with such opportunities. I also would like to acknowledge my colleagues in FIE-UCLV's Department of Electronics and Telecommunications, especially those in the radiocommunications group for their suggestions while I was working through the first few drafts of the manuscript. In addition, my thanks go to the editors and reviewers at Artech House, the publisher of this work.

Last, but not least, I wish also to express my appreciation and love to my widespread family, particularly to my daughter Lisette and son Jorge Luis and to my lovely wife Minerva for her invaluable support, especially in the initial phases of the project.

1

Introduction to Satellite Broadcasting Systems

1.1 Introduction

From its commercial beginning (1964–1965) until the end of the 1970s, satellite communication systems were mainly oriented toward very limited markets whose major customers were national governments. With the rapid expansion of satellite products into a variety of useful fields, entrepreneurs and others have begun to look upon space as an area with significant real estate value. Currently, it is widely recognized that satellite systems are contributing to advances in science and technology, the creation of new markets, progress in our information society, and improvements of our quality of life.

Traditionally, communication satellites have been used for point-to-point backbone communications such as international trunking telephony, while the fiber optic transmission system has come into use as a powerful means of backbone line communications. To take full advantage of the geometric capabilities of satellite systems, we must seek new areas for satellite. Emerging applications such as satellite digital TV, digital radio services, high-speed Internet communication services, and broadband multimedia services are now classified as direct satellite broadcasting systems.

In the years to come, satellites will play a major role in facilitating the information superhighway. Satellites are key to globalization because they are independent of distance; as a result, there is almost no relationship between cost and distance. Satellites will aid globalization through such avenues as advanced direct-to-home broadcasting and international Internet access.

Satellites will by no means replace wire-based transmission technologies, such as fiber optics. Instead, they will work in concert with them wherever they can deliver greater transmitting capability, lower cost, or more rapid deployment.

While the future cannot be accurately forecast, an adequate information infrastructure is undoubtedly an essential element for improving our quality of life. Satellite communication and broadcasting are keys to developing the global information superhighway. The satellite industry conducts business from a global perspective and contributes to the development of all humankind.

1.2   Architecture of a Satellite Broadcast System

A space segment and a ground segment (Figure 1.1) compose the satellite system for communications and broadcasting. The space segment contains the satellite and all ground facilities for the monitoring and control of the satellite (e.g., orbital position and adequate pointing to the coverage area on the Earth). Communication can be established between all ground stations located within the coverage area, also called the satellite footprint.

In an operational system, one or more in-orbit spare satellites usually back up the satellite in use. Many of the present communication satellites are in the geostationary orbit, and they are called *geostationary satellites*. The ground segment consists of all Earth stations directly connected to end-user equipment, such as transmitters and receivers. The uplink is the link between a transmitting Earth station and the receiver section of the satellite. The operating frequency band in the uplink (or feeder link) is denoted by f_U. The downlink is the link between the transmitting section of the satellite and a receiver Earth station, and f_D denotes the operating frequency band. The uplink and downlink frequency bands are commonly denoted by f_U/f_D in gigahertz.

With few exceptions, the uplink frequency band uses a higher operating frequency than the downlink frequency band, that is,

$$f_U > f_D, \text{GHz} \tag{1.1}$$

The reasons for this are listed as follows:

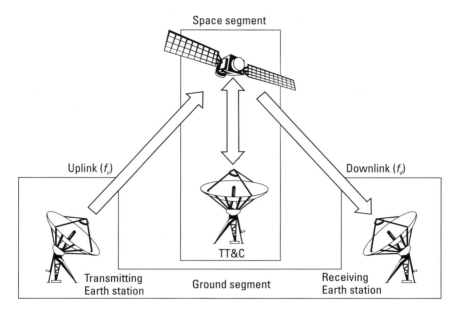

Figure 1.1 Satellite broadcast system elements.

- To avoid interference between the uplink and downlink;
- Because the satellite has a limited RF power output and transmission loss diminishes as frequency decreases.

The space segment uses a communication system other than the uplink and the downlink, between the satellite and *telemetry, tracking, and command* (TT&C) ground stations.

The design of a satellite network for direct broadcasting is based on whether the service to be provided is video-, audio-, or data-based. The objective of the design is to meet the desired *quality of service* (QoS) within all system constraints, such as the cost and the state of the art of the technology.

1.3 Radioelectrical Regions

The *International Telecommunication Union* (ITU) was formed on May 17, 1865, and became a United Nations agency in 1947. The ITU coordinates world meetings, known as *World Administrative Radio Conferences*

(WARCs), to discuss the future development of radiocommunication technologies, to assign *radio frequencies* (RFs) for new services, and to develop standards for the worldwide communication system. The ITU is responsible for allocating frequencies and orbital locations for domestic, regional, and international satellites.

There are three sectors within the ITU:

- The *ITU Development Sector* (ITU-D), which deals with development activities and mainly focuses on developing countries;

- The *ITU Radiocommunications Sector* (ITU-R), which develops RF allocations;

- The *ITU Telecommunications Sector* (ITU-T), which is responsible for the evaluation and standardization of all aspects of telecommunications, such as equipment interfaces and network communications.

Satellite communications and broadcasting are affected by all three ITU sectors. The ITU-R consists of 11 study groups, of which at least seven are closely related to satellite systems.

The ITU has divided the globe into three radioelectrical regions (Figure 1.2):

- ITU-1: Europe, Africa, and the Middle East (35° E to 56° W);

- ITU-2: America (57° W to 156° W);

- ITU-3: Asia, the South Pacific, and the Indian Subcontinent.

The radioelectrical regions (see Figure 1.2) reflect a kind of political and economical subdivision of the world for the purposes of allocating RF spectrum to ITU members. At WARC meetings, all regions are merged into one.

Radio regulations issued by the ITU specify detailed frequencies for various services and methods to avoid excessive interference between users. Although individual governments regulate the use of radio systems within their territory, international agreements are necessary to harmonize the use of radio on an international basis.

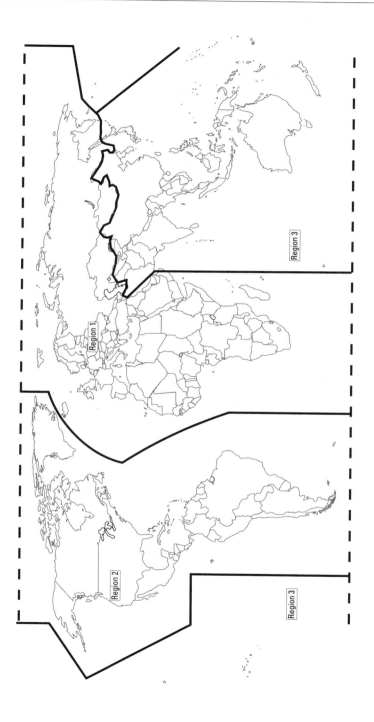

Figure 1.2 Radioelectrical regions.

1.4   Basic Satellite Services

The basic services for communication and broadcast are described as follows:

- *Fixed satellite services (FSSs):* Point-to-point transmissions not intended for direct public reception. FSSs are commonly network-oriented services where end users are transparent to satellite connection. A typical application is the intercontinental backbone network for telephony, data, and TV program exchange.

- *Broadcast satellite services (BSSs):* User-oriented services where the end user makes the satellite connection. Typical applications are satellite TV, *digital audio broadcasting* (DAB) systems, and satellite Internet services.

- *Mobile satellite services (MSSs):* Services that allow communication between mobile and fixed terminals. They are classified as *land mobile satellite services* (LMSSs), *aeronautical mobile satellite services* (AMSSs), and *maritime mobile satellite services* (MMSSs).

1.5   Satellite Frequency Bands

Satellite systems operate at frequencies ranging from 1 GHz to above 30 GHz—that is, using the ITU denomination, between the *ultrahigh-frequency* (UHF) and *extremely high-frequency* (EHF) bands. The basic designation for these bands is outlined in Table 1.1. For typical broadcast applications, satellite systems operate in Ku, C, L, and S bands.

Table 1.1
Satellite Frequency Bands

Band	Designation (GHz)	Typical Service
L	1–2	BSS (sound), MSS
S	2–4	MSS, BSS (sound)
C	(6/4)	FSS, BSS (TV)
X	7–8	Military, FSS
Ku	(14/11),(14/12),(17/12)	FSS, BSS (TV)
Ka	(30/20)	FSS, BSS, MSS

The L and S bands are particularly effective for providing rapid communication by means of mobile and transportable Earth stations.

The C band (6/4 GHz) is the major frequency band for commercial satellite communications, even as the Ku band has come into greater use. The principal advantage of C band is the low level of sky noise power; in addition, it is not practically affected by rain influence, and equipment technology and availability are positive factors. Today, the equipment has been made very inexpensive by the high-volume production for the global market. The main drawback of this band is the large antenna size required for user-oriented applications.

The Ku band became a very interesting frequency band for satellite TV services using inexpensive terminals with small-diameter receiving antennas. This band is divided into three segments: 14/11-GHz (FSS, region 1), 14/12-GHz (FSS, region 2), and 17/12-GHz (BSS). The Ku band is subject to rain attenuation, which can be overcome with a fade margin. This rain attenuation tends to reduce some of the benefits of the higher satellite power permitted by international regulations. It may be possible that in regions with the highest rainfall, the Ku band would be less practical than the C band.

The Ka band (30/20-GHz and about 2.5 GHz of available spectrum) is intended for commercial applications in cases where the C and Ku bands are fully occupied for incorporating new services.

The frequencies above 40 GHz (the Q and V bands) are now being evaluated for satellite communication applications because of the wide bandwidth and ability to transmit through very narrow spot beams. During heavy rainstorms the atmosphere will appear to be nearly opaque at these frequencies, so applications must allow for some periods of total outage.

1.6 Satellite Operators

A satellite operator provides satellite service to a specific coverage area, and it is the owner of the orbital location and the associated radio channels. A satellite operator must have relations with the following sectors:

- Satellite manufacturers, launch services, and insurance companies;

- Ground terminal equipment manufacturers and providers;

- Service companies [dealers and *conditional access* (CA)].

These relationships are depicted in the typical business environment in Figure 1.3.

A satellite operator, as a service provider, must provide a service with the following characteristics:

1. *Reliability:* The satellite industry ensures that satellites are designed, built, and launched with the highest level of reliability possible. One way to increase reliability is to use in-orbit spares, either with spare units on the spacecraft or with entire spare spacecraft.

2. *Transmission capacity:* This is achieved by increasing the number of transponders and connectivity through additional satellites and an increased emphasis on digital compression to squeeze more signals into each transponder bandwidth.

3. *Service:* Satellite operators are now offering more than just-in-time bookings for transmissions, for example, production and postproduction facilities for broadcasters and uplink facilities.

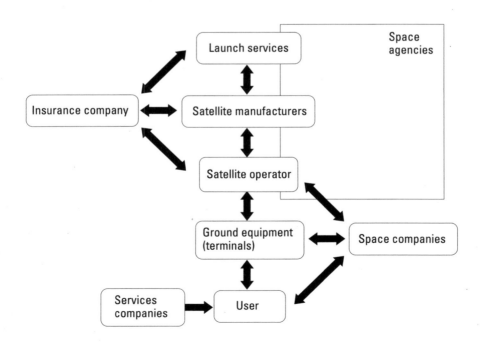

Figure 1.3 Satellite operator environment.

4. *Price:* Given the current demand for space segment and the cost of launching new satellites, it is unlikely that transponder costs will drop any time soon. A common price for a typical Ku-band transponder is about $200,000 per month. This can vary significantly with demand, supply, and contract terms.

According to the intended service, satellite operators can be classified as global operators (PanAmSat/Hughes, Intelsat, and Inmarsat), regional operators (SES-ASTRA and Eutelsat), and domestic operators (DirecTV, EchoStar, Hispasat, and Brazilsat).

1.7 Direct Broadcast Satellite Services

Satellite TV systems represent, by far, the most important satellite service today, with an expected annual average growth of 19% (between 1998 and 2002) [1]. These systems all use geostationary satellites and can be classified according to satellite power output as follows [2]:

- Low-power systems (C band; FSS: 6/4 GHz);

- Medium-power systems (Ku band; FSS: 14/12; 14/11 GHz);

- High-power systems or direct broadcast by satellite (DBS) (Ku band; BSS: 18/12 GHz).

The main features of these systems are summarized in Table 1.2 for region 2.

Satellites provide broadcast transmission in the fullest sense of the word, since the antenna footprints can be made to cover large areas of the Earth. The idea of using satellites to provide direct transmissions to the home has been around for many years [3].

C-band low-power systems have been very popular in North America. They have a standard frequency assignment for all orbital locations (24 transponders) and are not affected by rain. The C band is a low-noise frequency band, and it was necessary to develop very *low-noise amplifiers* (LNAs) and downconverters (LNBs) in the receiver ground terminal. The most important drawbacks of this band are its use of large antennas (2–4m) and an antenna tracking system (actuator) to tune several satellites in the geostationary orbit. These systems were the pioneers of analog satellite TV direct to users.

Table 1.2
Main Characteristics of Satellite TV Systems in Region 2

Parameter	HPS	MPS	LPS
Frequency band	Ku	Ku	C
Downlink (GHz)	12.2–12.7	11.7–12.2	3.7–4,2
Uplink (GHz)	17.3–17.8	14–14.5	5,925–6,425
Satellite service	BSS	FSS	FSS
Terrestrial interference	No	No	Yes
Minimum separation between adjacent satellites (degrees)	9°	2°	2°–3°
Satellite EIRP range (dBW)	51–60	40–48	33–38

DBS systems were first proposed at WARC'77 (1977) for regions 1 and 3 [4]. The committee developed technical specifications based on extrapolations of technology available at that time. Under the WARC'77 plan, each country in Europe was allocated up to five channels in either the lower subband (11.7–12.1 GHz) or upper subband (12.1–12.5 GHz) of the whole broadcast satellite service band. The WARC'77 plan also designated four orbital positions for European DBS operations (37°, 31°, and 19° W and 5° E). It also established 6° of orbital separation between DBS systems to minimize potential interference between adjacent satellites. Also, orthogonal circular polarizations were adopted to allow reuse of the limited spectrum available and to give a good isolation margin between adjacent DBS satellite systems using identical transponder frequencies. For example, the United Kingdom was assigned channels 04, 08, 12, 16, and 20, all with *right-hand circular polarization* (RHCP) receiving from the 31° W orbital position, while Andorra was assigned the same channels but with opposite *left-hand circular polarization* (LHCP) receiving from the 37° W orbital position. Circular polarization systems have the advantage of not requiring polarization orientation, which simplifies the installation of home receivers because the feed does not have to be rotated after the antenna is aligned in azimuth and elevation. However, circular polarization systems suffer more from depolarization during heavy rain than linear polarization systems. Some years later, the 1983 Regional Administrative Radio Conference (RARC'83) [5] did the same as WARC'77 for region 2 (America). The RARC'83 plan was based on the following essential aspects:

- A channel in the 12.2–12.7-GHz Ku band;

- An orbital position in the geostationary orbit to each country within the region and a minimum spacing of 9° of orbital separation between satellite systems serving adjacent or overlapping geographical areas;

- A description of the transmitting antenna beam of an elliptical cross section that just covers the service area in question;

- The use of RHCP or LHCP;

- The maximum allowable foresight power [*equivalent isotropic radiated power* (EIRP)] for that assignment.

The downlink frequency bands used for DBS systems vary from region to region throughout the world, but all use the Ku, or 12-GHz band. There are a number of reasons for the use of the higher frequency Ku band (rather than the C band) for DBS services. One major reason is the availability of frequencies at the higher frequencies. Also, the Ku band is not a designated band for terrestrial microwave systems, and so the interference problems associated with the C band do not arise. A feature readily observed with Ku-band reception is the comparatively small antennas that can be used. The antennas commonly utilize parabolic reflectors ("dishes"), and they are around 50 to 60 cm in diameter, compared to about 3m for C-band antennas. The smaller antenna size at the receiver is a direct result of the higher power available from the geostationary DBS satellites operating at Ku-band frequencies, compared to the C-band satellites.

RARC'83 developed a frequency-polarization scheme for the downlink and feeder-link bands. This scheme provides 32 overlapping radio channels of a 24-MHz bandwidth with adjacent channels cross-polarized (Figure 1.4). The channel center frequencies for the downlink band are given by

$$f_n = 12.224 + (n - 1) \cdot 0.01458, \text{GHz} \qquad (1.2)$$

where $n = 1 \ldots 32$.

RARC'83 defined two polarization plans: plan A (odd channels RHCP, even channels LHCP) and plan B (odd channels LHCP, even

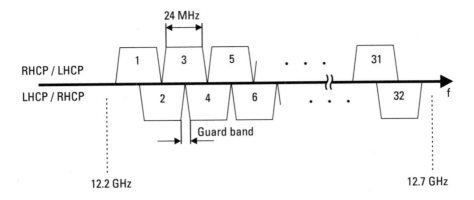

Figure 1.4 DBS channelization plan for region 2 (RARC'83).

channels RHCP). Cross polarization can add further protection as can frequency staggering. Since polarization plan A was used for nearly all orbital position assignments, adjacent satellites have larger geostationary arc separations than in WARC'77 (9° versus 6°) to provide more orbital interference protection. The United States—and most other countries having full 32-channel assignments (Argentina, Brazil, Canada, and Mexico)—reserved its right to use either sense of polarization on any channel.

System parameters defined in WARC'77 and RARC'83 for use in planning are shown in Table 1.3. This was a structured planning effort done by experts, but it does not take into account the technology evolution.

The advances in technology that immediately followed soon showed that the plans of WARC'77 and RARC'83 were impossible to implement and that they were not economically viable (for instance, the predicted maximum satellite's power output was around 60 dBW). By the late 1980s, satellite TV operators in the United States and Europe began broadcasting their signals using satellite frequencies and orbital positions that the ITU had originally assigned for FSS operators in the Ku band (14/11 GHz and 14/12 GHz) with denser frequency plans than that given by (1.2). SES-ASTRA was a typical satellite operator in Europe during this period. In the United States, C-band operators (e.g., Hughes-Galaxy) began to bring services to cable TV operators and to a market segment of about 4 million *direct-to-home* (DTH) terminals.

Although DBS systems were initially planned as analog systems (FM-NTSC, FM-PAL, and FM-SECAM standards), up-to-date DBS systems are digital. Digital video compression for the broadcasters along with

Table 1.3

DBS Systems Planned in WARC'77 and RARC'83

Parameter	WARC'77	RARC'83
Downlink frequency band (GHz)	11.7–12.5	12.2–12.7
Space segment bandwith (MHz)	800	500
Characteristics of radio channels:		
Number	40	32
Bandwith (MHz)	27	24
Adjacent radio channel separation	19,18	14,58
Guard bands (MHz), lower and upper	14&12	12&12
Polarization	circular	circular
PFD (99% worst month, coverage area boundary, dBW/m^2)	−103	−107
Receiver antenna diameter (m)	0.9	1
Receiver antenna −3-dB bandwith (degrees) (BW°)	2	1.7
CCI (dB)	31	29
ACI (dB)	15	13.6
G/T (dB.K)	6	10
C/N (dB, 99% worst month)	14	14
Modulation	FM (1 video + 1 sound	FM (1 video + 2 sound)
Minimum cofrequency adjacent satellite separation (degrees)	6°	9°
Envelope radiation pattern (normalized) for receiver antenna (copolar) (dB)	8, 5+25 log (θ/BW°)	14+25 log (θ/BW°)
Cross-polar isolation for receiver antenna (dB) (boresight)	−25 to −20	−25 to −20
Rain attenuation prediction method	ITU-R 5-zone	CPM model less 12% in rain attenuation in dB
Satellite station-keeping tolerance (degrees)	± 0.1 ns/ew	± 0.1 ew
Satellite antenna depointing tolerance and rotation (degrees)	0.1 /± 2	0.1 /±1

consistent picture and sound quality for customers are the driving forces behind the change to digital in satellite DTH TV program distribution. These digital television signals are transmitted in a format that significantly reduces the amount of frequency bandwidth required without substantially degrading the quality of the received pictures and sound. The introduction of compression technology is causing a notable decline in the operational costs of TV service providers. In 1994, Hughes-DirecTV opened a new epoch in the evolution of DBS systems (also called the "true" DBS) using compressed digital TV services for home terminals on a wide scale for the first time. The result has been a global explosion in the number of new satellite-delivered DBS TV services.

1.8 Concluding Remarks

One of the satellite's main advantages is its wide area coverage capability. In broadcasting systems, the downlink signal is available everywhere within the satellite's footprint. That capability will continue to be attractive for video, audio, and data delivery purposes. In fact, the 1990s saw the introduction of true DBS service, building on the foundation of medium- and high-power nonregenerative geostationary satellites and digital compression technologies. C-band TV service offerings from cable TV operators have given way to more attractive Ku-band services, which allow smaller antennas on receiving home terminals. The ability to deliver over 200 NTSC and PAL television programs from one orbit position is in place in many countries.

Another strong potential area for DBS systems is the delivery of *high-definition TV* (HDTV) programming. The digital standards for HDTV terrestrial broadcasts have been adopted in the United States [Advanced Television Standards Committee (ATSC)] and Europe [*digital video broadcasting-terrestrial* (DVB-T)]. HDTV has only recently been part of any satellite's published plans as a response to the plans of terrestrial transmissions. The greater capacity of satellites, compared to terrestrial channels, ensures that more bit rates are available to support HDTV both for single and multiple services. Terrestrial channels are particularly disadvantaged due to limited bandwidth (6 MHz and 8 MHz), but mostly because of poor channel quality caused by noise, interference, and unmanageable propagation events such as multipath. Whereas satellite transponders can deliver bit rates of several tens of megabits per second through a relatively stable channel (mainly disturbed by rain), the 6-MHz terrestrial channel is limited to a less than 20-Mbps net capacity and must also deal with multipath. Yet, the technology still appears to be bound by broadcasters' reluctance to make the required investments. The large corporations that have a proprietary system generally stimulate the incompatibility of systems, and this allows them to charge high prices.

The market of HDTV will represent, in the near future, billions of dollars in several economic sectors. The market's inherent economic and commercial values—and the national market economy integrity factors—spark the interest of governments and large corporations to control this technology. Their great preoccupation is justified by the fact that this technology is a consequence of the technological advances of semiconductor and display markets. Displays are used in practically all sectors of the electronic industry. Any equipment that employs a screen to communicate with the user may be affected by this new technology; this accelerates its technological aging process.

The technological and economical importance of HDTV does not diminish beyond the consumer electronic market borders. As a function of the nature of the related technologies and their spectrum of potential applications, it is possible to affirm that high-definition technology represents much more than a new product. It is a fundamental pole for the expansion of technological innovations to the entire electronic complex.

Nevertheless, while display sets are available, their high cost has impeded initial uptake. HDTV is an attractive idea, but its widespread adoption still rests on the assumption that enough viewers will purchase the much more expensive HDTV receivers when they become widely available. The basic question arises: Which will come first, the digital network or the millions of HDTV receivers? Add to that the challenge of making interesting programming available in this new format. Perhaps resistance to HDTV stems from the opposition of manufacturers to producing the right receivers before HDTV transmissions appear on the popular satellites.

The DBS platform, with its ability to deliver a high-speed digital time multiplex, also provides an important opportunity to deliver a variety of information services. Data broadcasting for Internet access can be introduced in the digital DBS systems now in use [6]. The same dish that provides over 200 programs of compressed TV and audio signals is also an effective way to distribute multimedia accessible with a Web browser.

There remains another application segment for satellite communication, that of providing DAB for a significant domestic or global market [7]. Experiments proved that digitized radio signals could be transmitted to vehicular terminals with a special L- or S-band antenna and receiver. One difficulty is that terrain blockage causes breaks in reception, something that probably would not be acceptable to the typical audience. This is overcome either by transmitting the same signal from multiple diversely placed satellites or through terrestrial broadcast towers to fill in gaps. Considerable effort has been expended in Europe on a DAB standard that would be flexible to short link outages. The FCC, in the United States, authorized DAB service by granting a license in 1997 for two satellite systems. One of these, XM Satellite Radio, uses two spacecraft in geostationary orbit and the other, Sirius Satellite Radio, has three spacecraft in elliptical high inclination orbit.

References

[1]   Taniguchi, I., "Vision of the Future Space Communications Industry," *Alcatel Telecommunications Review,* Fourth Quarter, 1999, pp. 241–243.

[2]   Reinhart, E. E., "Satellite Broadcasting and Distribution in the United States," *Telecommunication Journal,* Vol. 57, No. V1, June 1990, pp. 407–418.

[3]   Clarke, A. C., "Extraterrestrial Relays," *Wireless World,* October 1945, pp. 305–308.

[4]   Gould, R. G., and E. E. Reinhart, "The 1977 WARC on Broadcasting Satellites: Spectrum Management Aspects and Implications," *IEEE Trans. on Electromagnetic Compatibility,* Vol. EMC-19, No. 3, August 1977, pp. 171–178.

[5]   Reinhart, E. E., "An Introduction to the RARC'83 Plan for DBS in the Western Hemisphere," *IEEE Journal on Selected Areas in Communications,* Vol. SAC-3, No.1, January 1995, pp. 13–19.

[6]   Couet, J., D. Maugars, and D. Rouffet, "Satellites and Multimedia," *Alcatel Telecommunications Review,* Fourth Quarter, 1999, pp. 250–257.

[7]   Courseille, O., and P. Fournié, "WorldSpace: The World's First DAB Satellite Service," *Alcatel Telecommunications Review,* Second Quarter, 1997, pp. 102–108.

2

Satellite Platforms for Direct Broadcast Applications

2.1 Advantages of Geostationary Satellites

By far the majority (greater than 95%) of all satellite-broadcasting systems use geostationary satellites as platforms to provide the intended services. The main advantages are the following:

- The geostationary satellite is in a fixed position in the space related to any point on the Earth's surface; then, it is not necessary to use a tracking system on the receiver ground terminal, making it easier (and less expensive) to implement. Nonetheless, because of orbital perturbations, any geostationary satellite has residual motions around its nominal position in space. This induces antenna depointing and hence antenna-gain losses that may affect link performance.

- The geostationary satellite has a wide coverage area well-suited to broadcast applications. The transmission costs are independent of distance and the number of users receiving the signals.

- The geostationary satellite transmissions to Earth are done in nearly free-space conditions. Most of the results developed in communication theory in an *additive white Gaussian noise* (AWGN) reception environment could be well applied to satellite broadcasting systems engineering practice. Rain introduces the main transmission impairments in satellite broadcasting links operating over 10 GHz .

- The large propagation delay (250 ms) does not affect one-way broadcast transmissions from geostationary satellites. Interactive services (e.g., requests for programming packages and pay-per-view services) are implemented by way of complementary terrestrial networks, particularly telephone and cable networks.

- Geostationary satellite manufacturing and launch services now comprise a mature industry and use well-proven technology. This affords satellite operators a high confidence in the satellite broadcast industry and portends a safe market without foreseeable technical risks.

- Free-space loss between a geostationary satellite and any Earth station can be considered as a time-invariant loss for engineering purposes.

2.2 Geostationary Orbit

2.2.1 Elementary Concepts

The *geostationary orbit* (GEO) is a circular orbit in the equatorial plane, and it is concentric with the Earth's radius. A simplified geometric model is shown in Figure 2.1 to calculate the height (H) of the geostationary orbit. There are some simplifying assumptions:

- The Earth is considered to be a perfect sphere of radius R (6,378 km) and to be a body with a point mass M (5.974 · 10^{24} kg), located at its center.

- The geostationary satellite has a point mass m that ranges from 0.5 to 3 tons.

- There is no other force than Earth and the satellite in the dynamic analysis.

The satellite orbits around the Earth with a period T (23 hours, 56 minutes, and 4.1 seconds), the same as the rotational period of the Earth around its axis. This is the reason why the satellite is a "fixed" point in space related to an observer placed on any point on the Earth's surface. The

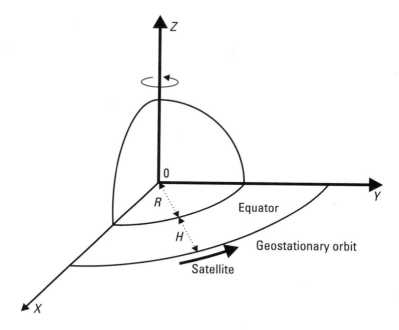

Figure 2.1 Simplified geometric model to calculate the height (*H*) of the geostationary orbit.

geostationary orbit is therefore unique and may be regarded as a limited natural resource.

There are two forces acting on satellites: the gravitational force F_g (attracting the satellite toward the Earth) and the centrifugal force F_c (ejecting the satellite from the Earth). To achieve dynamic equilibrium, it is possible to write

$$G_g \frac{mM}{(R+H)^2} = m \left(\frac{2\pi}{T} \right)^2 (R+H) \tag{2.1}$$

where G_g is the universal gravitation constant ($6.672 \cdot 10^{-11} \cdot \mathrm{Nm^2/kg^2}$). Using some algebraic manipulations from (2.1), it is possible to obtain

$$H = \sqrt[3]{G_g \cdot \frac{M \cdot T^2}{(2\pi)^2}} - R \tag{2.2}$$

Substituting numerical values in (2.2), one obtains H = 35,786 km, which is the altitude of the geostationary satellite above the equator.

2.2.2 Orbital Position

The orbital position (also known as the orbital slot) of a geostationary satellite is the satellite's location on the geostationary orbit. It is expressed as longitudinal degrees (west or east) of the subsatellite point on the Earth's equator. (The subsatellite point is the satellite's projection on the Earth's surface.) Each geostationary satellite is identified with one orbit position; for example, EchoStar I is placed in orbital position 119° W, while ASTRA 1F is placed at 19.2° E.

Several geostationary satellites may be copositioned in the same orbital position to expand capacity. It does not mean that two or more satellites are placed exactly in the same location in geostationary orbit. They are actually several tens or hundreds of kilometers apart from one another. This has very little effect on the figure that identifies the true orbital position for each copositioned satellite (see Example 2.1). This angular difference has not been taken into account in common practice (Figure 2.2). It should be noted that colocated satellites cannot use the same frequency bands.

Example 2.1

Satellites ASTRA 1A and ASTRA 1B are copositioned at 19.2° E. If the satellites are actually 20-km apart from each other in the geostationary orbit, calculate the true orbital position for ASTRA 1B, assuming that ASTRA 1A is exactly at 19.2° E and that it is longitudinally separate from ASTRA 1B leading in the direction of flight.

Solution. The angular difference between satellites ASTRA 1A and ASTRA 1B can be calculated as

$$\frac{20}{(6,378 + 35,786)} \cdot \frac{180°}{\pi} = 0.027°$$

Then, ASTRA 1B is placed exactly at 19.227° E.

2.2.3 Orbit Perturbations and Corrections

Geostationary satellites are not actually fixed bodies in space. There are some unbalanced forces that cause satellites to drift slowly away from their assigned

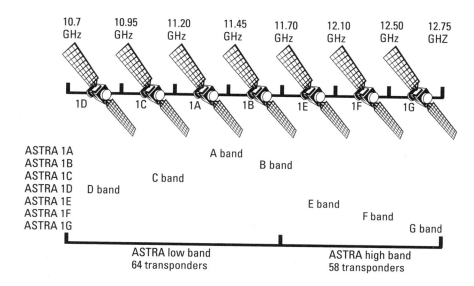

Figure 2.2 The ASTRA satellite constellation at 19.2° E.

orbital positions into figure-eight excursions with 24-hour periods that follow a path above or below the orbital plane. The main unbalanced forces stem from the following:

- The fact that the gravitational field of the Earth is not a symmetric one, since the Earth is not a perfect sphere and is rotationally asymmetric around its axis;

- The presence of the Sun, the Moon, and other bodies in outer space.

Ground controllers, commanding via TT&C stations, must periodically adjust satellite positions to counteract these forces. The process of maintaining a satellite within a preassigned tolerance in its orbital position (also called the slot or window) is known as *station keeping*. To maintain the satellite within the window, orbit corrections are achieved by applying velocity impulses to the satellite at a point in the orbit. These impulses are generated by activating the thrusters that are mounted on the satellite as part of the propulsion subsystem. The satellite can be kept operational as long as there is enough propellant for the thrusts; when no propellant is left, the satellite drifts in space out of control and comes to the end of its operational life. To avoid a possible collision with other geostationary satellites, its operators usually keep some amount of propellant to generate a final impulse that pulls it out of the GEO.

2.2.4 Sun Outage

The coincidence of the satellite and the Sun at the site of ground receiver terminal signifies that the Sun is viewed from the ground receiver terminal in the same direction as the satellite. As the antenna of ground receiver terminal is pointed toward the satellite, it also becomes pointed toward the Sun. The antenna catches the RF power coming from the hot Sun, and this additional input highly increases its antenna noise temperature.

As the satellites turn around the Earth, the Sun-ground terminal coincidence is a momentary event. It is predictable and actually happens twice a year for several minutes over a period of 5 or 6 days before the spring equinox (March 28) and after the autumn equinox (September 21) in the northern hemisphere; and after the spring equinox and before the autumn equinox for a ground terminal in the southern hemisphere. The practical outcome is a sudden degradation (perhaps an outage) in downlink performance because of the high rise of the antenna noise temperature.

2.2.5 Eclipses

During two, approximately 44-day seasons a year, centered around the vernal and autumn equinoxes, a geosynchronous spacecraft passes through the Earth's shadow daily. These transits of the shadow have a maximum of 72 minutes and have two consequences:

- The necessary use of backup batteries instead of solar cells as a source of energy;

- A resulting thermal shock that must be taken into consideration in the spacecraft design.

2.2.6 Doppler Effect

The Doppler effect is a change in the received frequency with respect to the transmitted frequency as a result of a nonzero velocity of the transmitter with respect to the receiver. When the satellite has a relative velocity of v meters per second along the line of sight, then the received carrier has a frequency shift given by v/λ where λ is the wavelength in free-space conditions. The frequency shift is positive as the satellite moves toward the receiver and negative as it goes away. Communications with geostationary satellites experience a small Doppler effect as a result of the movement of the satellite within its orbital window, and this has a very minor effect on broadcast applications.

2.2.7 Satellite Launchers

To launch a satellite into orbit, it is necessary to provide it with the appropriate velocity at a specific point of its trajectory in the plane of the orbit, starting from the launching base on the Earth's surface. This usually requires a launch vehicle (launcher) for the takeoff, and an onboard specific propulsion system.

With a geostationary satellite, the orbit aimed at is circular, in the equatorial plane, and an intermediate orbit called the *transfer orbit* attains it. Most conventional launchers inject the satellite into the transfer orbit at its perigee.

At this point, the launcher must give a velocity of 10,234 m/s to the satellite (for a perigee at 200 km). Then the satellite is left by itself and proceeds forward in the transfer orbit.

When arriving at the apogee of the transfer orbit, the satellite propulsion system is started, and an impulse is given to the satellite. This increases the velocity to the one required for the geostationary orbit (3,075 m/s). The satellite orbit is now circular, and the satellite has the proper altitude of 35,786 km from the equator.

The availability and cost of appropriate launch services are of critical importance. The growth of a private launch industry in the United States was promoted by a presidential policy statement in 1986, which followed the Challenger accident. As a consequence, commercial satellite launches via the Space Shuttle were terminated, and NASA programs for launching vehicles of the Shuttle class were eliminated. Since that time, commercial launches have been used almost exclusively for geostationary communications satellites, and a healthy, competitive launch industry exists and is gaining momentum. A new generation of powerful launch vehicles is making it possible for satellites to carry a greater number of transponders than ever before. At present, Arianespace (France, Ariane), International Launch Services (the U.S. Atlas/Centaur and Russian Proton), and Boeing (the United States' Delta and the Ukraine's Zenith Sea Launch) dominate the market. In addition, new launch services have appeared, including China's Long March and the Russian Soyuz.

2.3 Architecture of Geostationary Satellites

The architecture of a geostationary satellite comprises the bus (also known as the spacecraft or platform) and the communication payload. These are shown in Figure 2.3.

For a communication satellite, there are four main considerations:

- The kind of service to be provided (a DTH broadcast or a mixture of services);

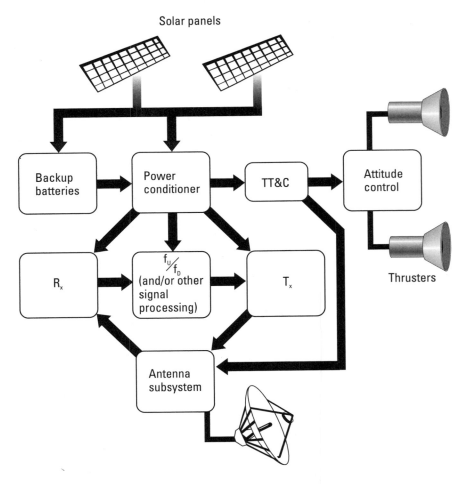

Figure 2.3 Geostationary satellite subsystems.

- The communication capacity—that is, the transponder bit rate with a given satellite *effective isotropic radiated power* (EIRP);
- The coverage area with the given satellite EIRP;
- The technological constraints (e.g., the launcher capability and the practical limit to the satellite's fuel storage capability).

The design of a communication satellite begins with a synthesis of a baseline spacecraft design, meeting all technical requirements such as EIRP and coverage. The synthesis process provides useful parameters such as the size and weight of the communication satellite. This is followed by the optimization of this basic design, taking into consideration technical constraints and costs. For instance, the assumed spot beam in the initial design may be too complex to implement, increasing the cost beyond the objective; therefore the constraints on the coverage area may have to be relaxed. A spacecraft, even one of a simple recurring design, is a highly engineered system for which extensive effort has been put into requirement definition, design, optimization, and validation.

The typical cost of manufacturing and launching a geostationary satellite is about $200 million. The cost of the communication payload is only about 20% of the total cost. The remaining cost is dedicated to the bus, testing, and launch. Launch costs generally include insurance, as roughly 10% of all launches end in failure before the intended start of service.

The communication payload comprises the satellite antennas and the microwave electronic equipment for processing the uplink carrier and sending it back to Earth. The bus allows the proper function (mission) of the communication payload.

2.3.1 Bus

All buses have seven main functions:

- *Structure:* Provides the satellite's mechanical integrity and supports all of its equipment. Low mass, rigidity, and carrying capacity are its main characteristics.
- *Thermal control:* Ensures an acceptable temperature range for the equipment by regulating energy exchanges with the space environment. In space, external exchanges are exclusively by radiation through the vacuum. Internal heat exchange within the satellite may be by conduction, convective heat pipes, and radiation.

- *Propulsion:* Correcting the trajectory after launch and changing or maintaining orbital and attitude characteristics during the mission require a propulsion system to supply the necessary impulses and velocity adjustments. Technology is moving away from chemical propulsion (propellants providing the energy that is transformed to thrust in jets) to electric propulsion methods (where some or all of the required energy is externally supplied) such as ion propulsion.

- *Electrical power supply:* Solar energy is collected by solar arrays (generators) and then conditioned for output to the equipment or storage in batteries for use during eclipse phases.

- *Attitude and orbit control:* Pointing antennas entails rotating the satellite around its center of mass with an accuracy ranging from 1° to 1 second. Maintaining orientation and keeping control during maneuvers, particularly orbit corrections, is the role of this complex function, which uses sensors, actuators, and an onboard computer. Generally attitude control subsystem electronics commands the propulsion subsystem, although there is usually (backup) direct ground commanding too.

- *Onboard management:* A satellite is always under control, either via transmissions to and from an Earth station (receipt of remote control signals and transmission of telemetry information) or through automatic control when out of contact with the Earth. It cannot be repaired once it is in geosynchronous orbit; this self-evident truth drives all the engineering talent dedicated to the architecture and choice of qualified components designed to guarantee a very high probability that the satellite will operate without problems. There is no room for error, even more so for a bus that is required to be as trouble-free and reliable as possible. Naturally, the user's main interest is the payload used for the mission.

- *Stabilization method:* The methods are body spin–stabilized satellites (low-power—e.g., HS 376) and three-axis–stabilized or body-stabilized (high-power—e.g., HS 601).

The competitive nature and stakes involved in international competition often focus on the bus trade names used by the major satellite manufacturers. This competitiveness is based on criteria such as cost, carrying capacity (mass and power), and launch qualification for various launch vehicles. The certified lifetime in orbit is of great importance and strongly influences the confidence of potential customers. Continuity, based on reusing what works well, means longer production runs and lower production costs. Nevertheless, a bus must remain flexible within its class of application to be able to meet the particular requirements of each mission and to enable the interfaces to be adapted to new payloads and new launchers. Taking account of these technical and commercial factors, the satellite industry has introduced a product-line approach for producing satellite buses and controlling the way they evolve. Leading manufacturers are now using ion propulsion systems, which considerably reduce the amount of station-keeping fuel that each satellite most carry into orbit. Satellite designers can therefore increase the mass of each satellite's communication payload to accommodate additional transponders.

2.3.2 Communication Payload

A communication payload is the system onboard the satellite that provides the link for the telecommunications signal path. The main functions of a nonregenerative communication payload are listed as follows:

- To receive and amplify the modulated carriers from the uplink;

- To change the uplink frequency band (f_U) to the downlink frequency band (f_D) of the received uplink-modulated carriers;

- To provide the RF output power required for the downlink-modulated carriers according to the satellite's frequency planning (channelization).

Two of the most important payload constraints are described as follows.

- *Payload mass:* Any increase in payload mass results in a heavier satellite and has a direct consequence on manufacture and launch cost.

- *Power consumption:* The satellite bus provides a certain amount of electrical power from its solar panels or batteries during eclipses; this power is mainly consumed by high-power amplifiers in the payload. The efficiency of the high-power amplifiers critically determines the number of transmitting channels in the satellite. Then, each piece of equipment in the payload should be designed to minimize the overall power consumption.

Figure 2.4 shows the general architecture of the communication payload for a specific type of polarization. Usually, a diplexer is used to separate the receiving and transmission paths to use a single antenna system for received and transmitted modulated carriers. The receiver section encompasses a wideband (passband) amplifier and an LNA. It follows the frequency downconverter, which also may be included in the satellite receiver section. The *input multiplexer* (IMUX) splits the incoming modulated carriers into independent groups of one modulated carrier, as is usual for broadcast applications. Each modulated carrier is going to be amplified to the RF power level required for transmission in the *high-power amplifier* (HPA) section. The output power amplifier is generally a *traveling wave tube amplifier* (TWTA) or a *solid-state power amplifier* (SSPA).

The TWTA operates by interaction between an electron beam and the radio wave. The interaction leads to a slowing of the electrons, which give up their kinetic energy. Typical values of the characteristic of the TWTA are listed as follows:

- Power at saturation: From 5W to 100–200W;

- The dc to RF efficiency including electronic power conditioning: 40–70%;

- Gain at saturation: Approximately 55 dB.

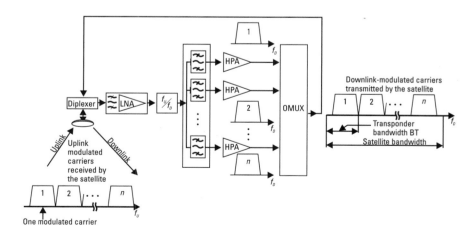

Figure 2.4 Communication payload.

The SSPA uses *field effect transistors* (FETs). The available power and operating frequency are improving as technology progresses. The output power required is achieved by connecting transistors in parallel in the final stages. The main characteristics are listed as follows:

- Power: 30W at 12 GHz, with more possible at lower frequencies;
- Efficiency: 20–35%;
- Gain at saturation: Approximately 50 dB.

The HPA used on the transmitter section of the communication payload operates in saturation (Figure 2.5) for broadcast applications to obtain maximum output power. This is a highly nonlinear region, and it is necessary to amplify only one modulated carrier to avoid intermodulation products.

The absolute time delay between input and output signals at a fixed input level in a TWTA is generally not significant. However, at higher input levels, where more of the electron-beam energy is converted to output power, the average beam velocity is reduced, and therefore the delay time is increased. Since phase delay is directly proportional to time delay, this results in a phase shift, which varies with input level. Thus, if the input signal power level changes, phase modulation will result in a process called *AM/PM conversion*. The slope of the phase shift characteristic gives the phase modulation coefficient, in degrees per decibel.

TWTAs are still preferred for satellite broadcast applications because their output power is higher than that of SSPAs. SSPAs are lighter than TWTAs, and they need a lower power supply to deliver the same output power. The channelization used in the satellite's bandwidth provides a means for efficient amplification of the modulated carriers, limiting the effects of the nonlinear properties of the HPAs, which generate the intermodulation products. It also provides a more flexible operation of the satellite. The output of each HPA is then combined in the *output multiplexer* (OMUX) and fed to the transmitting antenna.

A *transponder* (transmitter-responder) is the series of interconnected hardware units in the payload (e.g., amplifiers, filters, and waveguides) that forms a single communications channel between the receive and transmit satellite antennas to process each modulated carrier. Some of these units are common to a number of modulated carriers. In broadcast applications, a transponder is commonly used to identify one transmission channel from a satellite. The polarization used in communication satellites is either linear or circular:

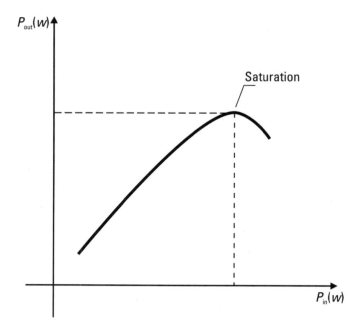

Figure 2.5 Power characteristics for single-carrier operation.

- Linear polarization: horizontal (H) and vertical (V), often with an offset angle;

- Circular polarization: RHCP (clockwise, viewed in the direction from the satellite to the Earth) and LHCP (counterclockwise).

As mentioned in Chapter 1, orthogonal polarization is commonly used with frequency reuse to increase the capacity of satellites in terms of the number of transponders for the fixed satellite bandwidth (for instance, 500 MHz), if sufficient power is available. A typical example is the configuration of twenty-four 36-MHz transponders used in C-band satellites in region 2. This technique effectively doubles the number of channels that can be provided by a satellite since two channels can share the same frequency bandwidth, given that they have opposite polarizations. As it is known, a wave with a given polarization does not transfer RF power to a receiving antenna that is orthogonally polarized to the wave. Thus, a vertically polarized receiving antenna does not receive anything from a horizontally polarized wave, and an LHCP-polarized receiving antenna does not receive anything from an

RHCP-polarized wave. Satellite operators use this orthogonal property of antenna polarization to allow RF carriers to overlap in frequency without interference (frequency reuse). Because any receiving antenna used in satellite communications has a cross-polar radiation pattern, only a practical 20–30-dB-isolation level between two selected orthogonal senses of polarization can usually be obtained [1]. As a further technique of isolation protection, satellite transponder center frequencies typically, but not always, are staggered from one set of like-polarization transponders to the other. This half-transponder offset configuration places the highest RF power region of a transponder into the low-RF power region of the opposite polarization, and vice versa. It is important to underline that copolarized channels denote channels of the same polarization and that cross-polarized channels are of opposite polarization.

It is usually observed that some medium-power DBS satellites do not use frequency reuse with orthogonal polarizations. For example, EchoStar I (119° W) has sixteen 24-MHz transponders with RHCP and that the colocated EchoStar II (the same orbital position as EchoStar I) has sixteen 24-MHz transponders with LHCP. Both satellites implement the full 32-channel assignment for the 119° W orbital position. There is no alternative to expand the satellite's channel capability, because this is mainly limited by power supply considerations. This approach has the positive aspect of protecting the service against the failure of a single satellite.

Adjacent DBS satellites in the geostationary orbit, due to their high RF power output, usually have opposite polarizations to diminish interference between signals on the downlink side, as planned in WARC'77. This is not the case in the RARC'83 plan, where adjacent satellites usually use copolarized frequency channels. For example, orbital positions 110° W and 119° W, today corresponding to EchoStar I-II and EchoStar V, respectively, have both assigned the full 32-channel arrangement with the same polarization plan A.

Example 2.2

Determine the maximum number of TV channels that a satellite, using a zone beam of gain 32.2 dBi, can provide given the following parameters:

EIRP/channel: 53 dBW (BSS Ku band);

Electric power: 4.3 kW;

The dc-to-RF conversion efficiency: 50%;

90% of the power is used by the RF power amplifiers.

Solution.

> Total RF power available: $0.5 \cdot 0.9 \cdot 4{,}300 = 1{,}935$W;
>
> TWTA power output: $10^{(53 - 32.2)/10} = 120.2$W;
>
> Number of TWTA/channels (maximum): $\dfrac{1{,}935}{120.2} = 16$;
>
> Maximum number of transponders: 16.

The transponders used in the communication payload architecture shown in Figure 2.4 are also called bent-pipe transponders. This kind of transponder is just like an analog repeater in the sky and does no signal reformatting.

2.3.3 Single-Beam and Multibeam Communication Payload

The communication architecture payload described in Figure 2.4 allows the implementation of a single-beam satellite network, as shown in Figure 2.6. There are several general classes of (single) beams, according to their coverage area:

- *Spot and zone beams:* They cover less than 10% of the Earth's surface. The satellite power output can be more concentrated and allows the use of small receiving antennas—for direct broadcasting.

- *Hemisphere beam:* It covers up to 20% of the Earth's surface, and the satellite's power output is 50% lower (–3 dB) than those transmitted by spot and zone beams. Therefore, it is necessary to use larger receiving antennas than for spot and zone beams.

- *Global beam:* It covers 42.4% of the Earth's surface, and large receiving antennas must be used to adequately detect audiovisual and data broadcasts.

It is well known that the same satellites broadcast toward two or more different geographic areas to optimize the satellite output power. For example, in region 2, Galaxy V (1.25° W) has one zone beam to the contiguous United States (CONUS) and two spot beams pointed to Puerto Rico and Hawaii, respectively. Hispasat 1 A (30°) region 1, has a single zone beam to Spain and Portugal and one spot beam to the Canary Islands. If it were to use a single wide coverage area zone beam to focus on the whole area, it would have been necessary to use a higher antenna gain in the receiver ground terminal because of the lower received satellite power. There must be, at least, as

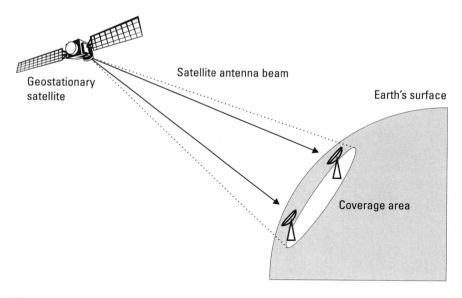

Figure 2.6 Single-beam satellite network.

many different onboard satellite antenna ports as there are different coverage areas and associated kinds of beams.

When a satellite network is used for broadcasting, it is necessary to make decisions between interconnecting a large number of receiving ground terminals within an intercontinental coverage area and large receiving antennas or achieving high-power output by means of a high satellite antenna gain with reduced coverage area and small receiving antennas.

The interconnected multibeam satellite network allows information exchange between continents or countries very far apart each (e.g., North America, Europe, and South America) from the interconnection of independent nonwide coverage area beams belonging to different continents or countries.

As indicated in Figure 2.7, the signals received in zone A are uplinked to the satellite and are downlinked to the same zone A along with the signals received from zone B. This situation can be achieved using a communication payload architecture as shown in Figure 2.8 by means of an interconnection matrix. This matrix is made of a set of interconnecting passband filters and other hardware that allows the modulated uplink carriers coming from a determined beam to be switched to another beam for the downlink. A very typical example is Intelsat K, which has three interconnected zone beams:

North America, Europe (both with up- and downlinks), and South America (only with downlink). The satellite antenna technology and the mass and complexity of satellites limit the number of beams.

2.3.4   Communication Payload with Regenerative Transponder

As was pointed out, a communication payload is nonregenerative (or transparent) when the modulated carrier is amplified and frequency-downconverted without being demodulated. A communication payload with regenerative transponders implies onboard demodulation of the uplink carrier, and then, the information signal is used to modulate a new downlink carrier. This kind of signal processing is well-suited to digital carriers.

The performance of a communication payload with a regenerative transponder is measured in terms of the overall *bit error rate* (BER) in the end-to-end satellite link. Let P_U and P_D be the bit error probability on the uplink and downlink, respectively. Assuming that the BER events in the uplink and downlink are statistically independent, then the probability of a bit not being in error (correct reception), or P_C, in the end-to-end satellite link is given by

$$P_C = (1 - P_U)(1 - P_D) \qquad (2.3)$$

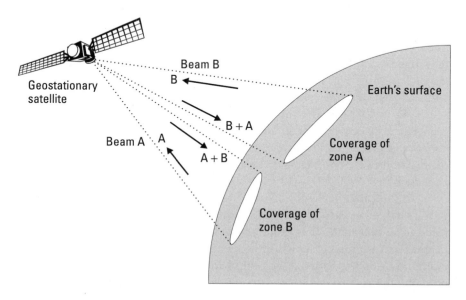

Figure 2.7 Multibeam satellite network.

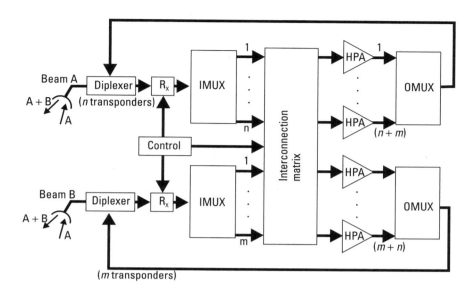

Figure 2.8 Communication payload for a multibeam (two-beam) satellite.

Therefore, the bit error probability in the end-to-end link is

$$P_B = 1 - P_C = P_U + P_D - P_U P_D \qquad (2.4)$$

Since P_U and P_D are much less than unity,

$$P_B = P_U + P_D \qquad (2.5)$$

Equation (2.5) illustrates the virtual independence between the uplink and the downlink. If $P_U \ll P_D$, then the overall link performance is dominated by the downlink.

Figure 2.9 shows a typical advanced communication payload using a regenerative transponder, also called a communication payload with *onboard processing* (OBP). Several low data rate bit streams at the output of demodulators can be combined into a *time division multiplex* (TDM) that modulates a single high data rate IF carrier and then travels to the frequency composer to generate the downlink carrier.

It should be pointed out that current commercial geostationary satellites do not use regenerative communication payloads but only transparent ones. NASA's *Advanced Communications Technology Satellite* (ACTS),

ITALSAT (Italy), Nilesat (Egypt), Hot Bird (Eutelsat, partially or several transponders) [2], and WorldSpace satellites use communications payloads with OBP.

2.4 Main Technical Characteristics of a Communication Satellite

The technical characteristics of any communication satellite used in broadcast applications largely determine the performance capabilities of each analog or digital service using that technological platform. The main technical characteristics include the following:

- The transmitting frequency bands (downlink) and polarization. For example, in the BSS service defined by ITU, the following frequency allocations: 11.7–12.1 GHz (lower band) and 12.1–12.5 GHz (upper band) for regions 1 and 3; 12.2–12.7 GHz for region 2 with orthogonal polarization (RHCP and LHCP).

- The satellites' orbital position and a basic overview. The fleet of EchoStar satellites may be used as an example:

  - EchoStar I: It was launched in 1992 and operational on March 4, 1996. The orbital location is 119° W; the coverage area includes CONUS, Southern Canada, and Northern Mexico; and it has sixteen 24-MHz transponders carrying about 100 digitally compressed programs.

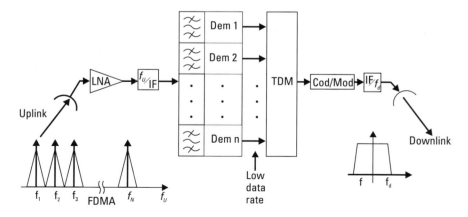

Figure 2.9 Communication payload with OBP.

- EchoStar II: It is a colocated satellite at 119° W and has the same characteristics as EchoStar I. Both satellites cover the full 32 channels allocated to this orbital position (RARC'83). It was launched on September 10, 1996.

- EchoStar III: It covers an area from the central (mountain) to eastern United States and is located at 61.5° W with a capacity of thirty-two 24-MHz transponders and 75 programs available. It was launched on October 5, 1997.

- EchoStar IV: It provides broadcast networks, international and niche channels, educational and business television, and data delivery applications. It is located at 148° W and covers CONUS with thirty-two 24-MHz transponders with 75 programs available. Originally intended for use in 119° W, it is at the 148° W slot, which has limited visibility from CONUS, because not all 32 transponders are available due to the nondeployment of one solar array and (unrelated) TWTA failures.

- EchoStar V: It was launched on September 23, 1999. The satellite is located at 110° W and is the Digital Sky Highway (DISH) Network's primary resource for its core services. The number of programs available is 150, using thirty-two 24-MHz transponders.

- EchoStar VI: It was launched on July 14, 2000. The satellite is located at 119° W. It covers CONUS, Hawaii, Alaska, and Puerto Rico, using thirty-two 24-MHz transponders with 150 programs available.

  The DISH Network has the capacity to offer over 500 digital video, data, and audio channels of programming, including local networks and HDTV to over 3 million customers (as of October 2000). This represents a total capacity of 4.42 Gbps.

- Footprint map. It is the main beam radiation pattern of the transmit antenna onboard a satellite broadcast modulated carrier toward a geographic region within the view of the spacecraft (coverage area). This radiation pattern is measured in terms of EIRP and is expressed in decibels relative to 1W. A footprint map is graphically represented by a set of isogain contour lines superimposed on the map of the geographic region served (Figure 2.10) [3].

The EIRP can be defined by the following equation:

$$\text{EIRP}(\text{dBW}) = P_T(\text{dBW}) + G_T(\text{dBi}) - L_C(\text{dB}) \qquad (2.6)$$

where P_T (dBW) is the HPA output power expressed in decibels per 1W, G_T is the transmit satellite antenna gain toward a specific direction on the coverage area, expressed in decibels relative to an isotropic source (dBi), and L_C is the coupling loss between the HPA output and the satellite transmit antenna input. The term EIRP comes from the word isotropic and means equal in all directions. EIRP identifies the power levels that would be received at any location if an antenna were radiating equally in all directions. This is so because the term transmit antenna gain (G_T) is "absorbed" in this definition and seems like an isotropic radiator $(G_T = 1)$. Equation (2.6) is associated for each transponder in the communication payload and the specific spacecraft transmit antenna gain to wherever the user receiving the signal happens to

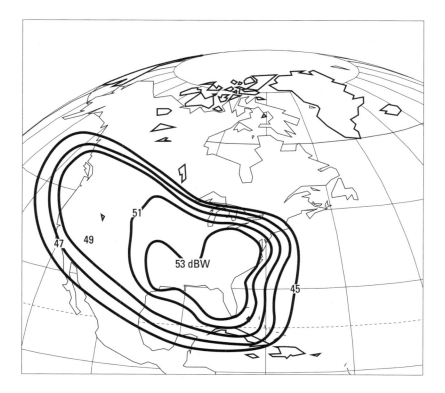

Figure 2.10 EIRP contours.

be. The spacecraft transmit antenna performance determines the shaping of the EIRP coverage.

- The figure of merit $(G/T)_S$ of the satellite receiving system, expressed in decibels per Kelvin, is proportional to the received uplink signal level versus the thermal noise floor level. This parameter, used for the uplink design, is the more fundamental of the two satellite receive sensitivity parameters [the other being the *saturated flux density* (SFD), described next].

- The SFD is the power flux density required at the onboard receiving antenna in order to produce the saturated performance at the transponder output and it is expressed in dBW/m^2. This parameter is needed by uplinkers because $(G/T)_S$ is a ratio only. One can see the need for specifying SFD by considering that two satellites with the same $(G/T)_S$ might have different front-end noise temperatures and need a different amount of power to saturate their HPAs, which need not have the same gains.

- Design life is an estimate of the satellite's expected operational lifetime and largely depends on the use of propellant to maintain the attitude and nominal orbital position (station keeping) and aging of solar cells. Up-to-date typical values are in the range of 12 to 15 years.

- Digital DTH transmission parameters are used to tune the digital *integrated receiver and decoder* (IRD) and includes the channel's digital transmission rate expressed in megasymbols per second, the code rate, and the CA system in use. For example, the DISH Network uses a symbol rate of 20 Msymb/second and a fixed FEC rate of 3/4 and the CA is Nagravision. All transponders in EchoStar satellites (EchoStar I, II, III, and IV) use the same DVB-S standard transmission parameters.

References

[1]   Claydon, B., "Introduction to Antennas," in *Satellite Communications Systems*, B. G. Evans (ed.), London: IEE, 1999, pp. 83–98.

[2]   Elia, C., and E. Colzi, "Skyplex: Distributed Uplink for Digital Television via Satellite," *IEEE Trans. on Broadcasting*, Vol. 42, No. 4, December 1996, pp. 305–313.

[3]   Baylin, F., *1998/2000 World Satellite Yearly*, Boulder, CO: Baylin Publications, 1998.

3

Satellite TV Link Analysis

3.1 Radio Transmission Equation Overview

Figure 3.1 shows the thermal noise model of a point-to-point radio transmission system, which can be used in a satellite channel that is mainly affected by attenuation and AWGN.

The *carrier-to-noise ratio* (C/N), defined as the quotient between the modulated carrier power at *intermediate frequency* (IF) and the corresponding noise power in the noise bandwidth of the IF amplifier (B), can be written as

$$C / N = \frac{P_T G_T G_R}{k T L_b L_a B} \tag{3.1}$$

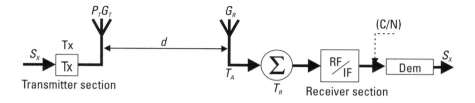

Figure 3.1 Thermal noise model of a point-to-point radio transmission system.

where:

P_T: Transmitter radiated power at carrier frequency, in watts (W);

G_T: Transmitting antenna gain;

T: System's noise temperature, in Kelvin (K);

L_b: Free-space loss.

L_a: Additional loss (relative to free-space loss);

k: Boltzmann's constant ($1.38 \cdot 10^{-23}$ dB.J/K).

The system's noise temperature T is defined as

$$T = T_A + T_R \tag{3.2}$$

where T_A is antenna noise temperature (external noise) and T_R is the receiver noise temperature (internal noise), both in Kelvin. (External and internal noise powers can be added because they are uncorrelated noise sources.)

The free-space loss is defined as

$$L_b = \left(\frac{4 \pi d}{\lambda} \right)^2 \tag{3.3}$$

where d is the distance (in meters) between the transmitting antenna and the receiving antenna, and λ is the wavelength (in meters) in the vacuum. The relationship between the carrier frequency f (in gigahertz) and wavelength λ (in meters) is

$$\lambda(\text{m}) = \frac{0.3}{f(\text{GHz})} \tag{3.4}$$

For satellite system applications, it is possible to introduce the following parameters:

- In the transmitting side:

$$\text{EIRP} = P_T G_T , \text{W} \tag{3.5}$$

- In the receiving side:

$$G / T = \frac{G_R}{T_A + T_R} ; \text{K}^{-1} \tag{3.6}$$

Now, using (3.5) and (3.6), it is possible to write (3.1) as

$$C / N = \frac{EIRP \cdot G / T}{k L_b L_a B}$$

(3.7)

Using decibels,

$$C / N(dB) = EIRP(dBW) + G / T(dB/K)$$
$$-L_b(dB) - L_a(dB) - B(dB.Hz) + 228.60(dBW / K.Hz)$$

(3.8)

Example 3.1

A transmitter feeds a power of 120W into an antenna that has a gain of 32 dBi. Calculate the EIRP in watts and decibels relative to 1W.

Solution.   Using (3.5), obtain

$$EIRP = 120 \cdot 10^{32/10} = 190.2 \text{ kW}$$

and

$$EIRP = 10 \log(120) + 32 = 52.8 \text{ dBW}$$

Example 3.2

A receiving system employs a 36-dBi parabolic antenna operating at 12 GHz. The antenna noise temperature is 50K, and the receiver front-end noise temperature is 110K. Calculate G/T in decibels per Kelvin.

Solution.   Using (3.6), obtain

$$G / T = 36 - 10 \log(50 + 110) = 13.96 \text{ dB/K}$$

3.2   Satellite Link Geometry

3.2.1   Slant Range

Figure 3.2 shows the basic geometric model of the satellite-ground terminal path.

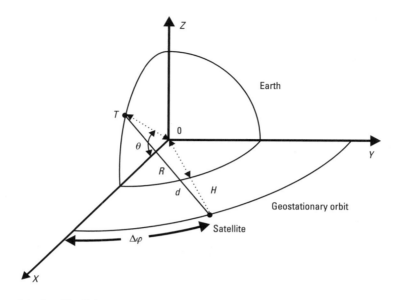

Figure 3.2 Satellite link geometry.

Let θ be the ground terminal (T) latitude, φ_T be the ground terminal longitude, and φ_S be the satellite's orbital position. North latitudes and east longitudes will be considered positive-signed; south latitudes and west longitudes are negative-signed. Define $\Delta\varphi$ as

$$\Delta\varphi = \varphi_S - \varphi_T \tag{3.9}$$

The rectangular coordinates of the ground terminal (point T on the ZX plane) are

$$R\cos\theta;\ 0;\ R\sin\theta$$

and of the satellite (point S on the XY plane) coordinates are

$$\left[(R+H)\cos\Delta\varphi;\ (R+H)\sin\Delta\varphi;\ 0\right]$$

The slant range d can be calculated using the well-known formula of distance between two points in the space knowing the coordinates of T and S. The end result is

$$d = \sqrt{(R+H)^2 + R^2 - 2R(R+H)\cos\Delta\varphi\cos\theta} \qquad (3.10)$$

Substituting the numerical values of R and H in (3.10), obtain

$$d = 4.264 \cdot 10^4 \sqrt{1 - 0.296\cos\Delta\varphi\cos\theta}, \text{km} \qquad (3.11)$$

Using (3.3), (3.4), and (3.11) and decibels, it is possible to obtain

$$L_b(\text{dB}) = 185 + 20\log f(\text{GHz}) + 10\log(1 - 0.296\cos\Delta\varphi\cos\theta) \;(3.12)$$

Example 3.3

A ground terminal located at 22° N, 80° W is receiving signals from geostationary satellite Galaxy V (125° W).
   Determine the free-space loss at 4 GHz (C band).

Solution.   Using (3.12),

$$L_b = 185 + 20\log(4) + 10\log[1 - 0.296\cos(-45°)\cos(22°)] = 196.1\,\text{dB}$$

Example 3.4

Repeat Example 3.3, but now the ground terminal receives a signal from EchoStar I (119° W) operating at 12.5 GHz (Ku band).

Solution.   Again, using (3.12), obtain the following result:

$$L_b = 185 + 20\log(12.5) + 10\log[1 - 0.296\cos(-39°)\cos(22°)] = 205.9\,\text{dB}$$

3.2.2   Elevation and Azimuth Angles

The elevation (EL°) and azimuth (AZ°) angles allow an antenna of a ground terminal to be accurately pointed to a satellite in the geostationary orbit (Figure 3.3).
   It can be shown (Appendix A) that the elevation angle is expressed as

$$\text{EL}° = \arctan\left(\frac{\cos\Delta\varphi\cos\theta - 0.1513}{\sqrt{1 - \cos^2\Delta\varphi\cos^2\theta}}\right) \qquad (3.13)$$

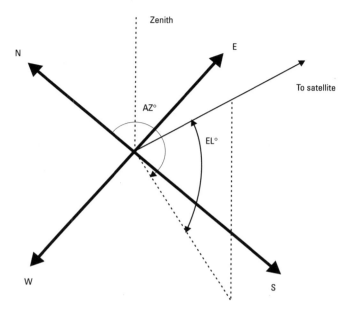

Figure 3.3 Elevation (EL°) and azimuth (AZ°) angles.

and the azimuth angle as

$$AZ° = 180° - \arctan\left(\frac{\tan \Delta\varphi}{\sin \theta}\right) \tag{3.14}$$

Example 3.5

Calculate the azimuth and elevation angles for a ground terminal located at 22° N, 80° W receiving digital TV from EchoStar 1 (119° W).

Solution.

$$EL° = \arctan\left(\frac{\cos(-39°)\cos(22°) - 0.1513}{\sqrt{1 - \cos^2(-39°)\cos^2(22°)}}\right) = 39.4°$$

$$AZ° = 180° - \arctan\left(\frac{\tan(-39°)}{\sin(22°)}\right) = 245.2°$$

Example 3.6

Allowing a 5° elevation angle at ground terminals, verify that the geostationary arc of 55–136° W covers CONUS.

Solution. Let us consider two locations in the United States: Washington, D.C., (39° N, 77° W) on the East Coast, and very near to Seattle (48° N, 125° W) on the West Coast.

Let

$$x = \cos \Delta\varphi \cdot \cos \theta$$

in the elevation angle formula. The variables latitude (θ) and relative longitude ($\Delta\phi$) are constrained by

$$0 \le |\theta| < 90°$$

$$0 \le |\Delta\varphi| < 90°$$

and then $0 < x \le 1$. For EL° = 5° and using the elevation angle formula, obtain

$$x^2 - 0.3x - 0.0047 = 0$$

with the only possible solution $x = 0.315$. The other one is neglected ($x = -0.015$) because of the physical constraints. Then,

$$\cos \Delta\varphi = \frac{0.315}{\cos \theta}$$

and, using the definition of $\Delta\phi$, it is possible to set

$$\varphi_S = \varphi_T \pm \arccos\left(\frac{0.315}{\cos \theta}\right)$$

From the location with coordinates 39° N, 77° W, the visible geostationary arc is

$$11° \, W \le \varphi_S \le 143° \, W$$

and it contains 55–136° W.

From the location with coordinates 48° N, 125° W, the visible geostationary arc is

$$50° \, W \le \varphi_S \le 185.66° \, W$$

and it also contains 55–136° W.

3.3 Pointing an Antenna Dish Using a Compass as a Reference

Once the azimuth angle has been calculated, if one wishes to point a dish using a magnetic compass, it is necessary to know the right direction of true north/south by using a compass and correcting for magnetic variation of the receive site. Points of equal magnetic variation on the Earth are represented with isogonal lines. When magnetic variation is zero, the contour is called the *agonic line*. Magnetic variation is added to the azimuth angle if the direction of magnetic north lies to the east of the true north and subtracted if it lies to the west. For example, if the azimuth angle is 153° and the magnetic variation of the receive site is 6° W, the compass bearing is 159°.

3.4 Overall RF Satellite Link Performance

The overall performance of the total satellite link depends on the uplink, downlink, transponder, and interference. Figure 3.4 shows a schematic model of a nonregenerative transponder including uplink and downlink. Only the effects of (thermal) noise will be considered.

The C/N for the uplink is defined as

$$(C/N)_U = \frac{P_U}{N_U} \tag{3.15}$$

In the same way, for the downlink

$$(C/N)_D = \frac{P_D}{N_D} \tag{3.16}$$

The overall noise power arriving to the demodulator input of the receiving ground terminal is

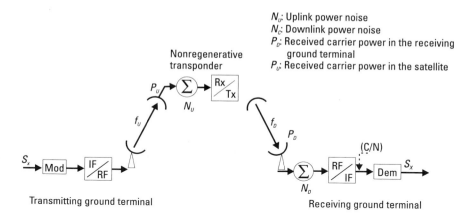

N_U: Uplink power noise
N_D: Downlink power noise
P_D: Received carrier power in the receiving ground terminal
P_U: Received carrier power in the satellite

Nonregenerative transponder

Transmitting ground terminal

Receiving ground terminal

Figure 3.4 Thermal noise model of an overall satellite link using a nonregenerative transponder.

$$N = N_D + \frac{N_U \cdot G_{TR}}{L_D} \qquad (3.17)$$

where G_{TR} represents the transponder gain and L_D represents the downlink transmission loss (including antenna gain).

The overall C/N (measured in the demodulator's input of the receiving ground terminal) is

$$C/N = \frac{P_D}{N} \qquad (3.18)$$

Taking the reciprocal of C/N in (3.18) and substituting (3.17), obtain

$$(C/N)^{-1} = \frac{N_D}{P_D} + \frac{N_U}{\left(\dfrac{P_D \cdot L_D}{G_{TR}}\right)} \qquad (3.19)$$

The term in parentheses in (3.19) is P_U; then,

$$(C/N)^{-1} = \frac{N_D}{P_D} + \frac{N_U}{P_U} \qquad (3.20)$$

Substituting (3.15) and (3.16) in (3.20), obtain

$$(C/N)^{-1} = (C/N)_U^{-1} + (C/N)_D^{-1} \qquad (3.21)$$

Note that this has exactly the same form as the equation for resistance of resistors in parallel.

It is possible to rearrange (3.21) in the following way:

$$(C/N)^{-1} = (C/N)_D^{-1}\left(1 + \frac{(C/N)_D}{(C/N)_U}\right) \qquad (3.22)$$

and then

$$(C/N)_D = C/N\left(1 + \frac{(C/N)_D}{(C/N)_U}\right) \qquad (3.23)$$

The term in brackets physically means the uplink noise contribution to overall noise in the satellite link. If this uplink noise contribution is denoted by ΔN_U, then

$$\Delta N_U = \left(1 + \frac{(C/N)_D}{(C/N)_U}\right) \qquad (3.24)$$

Substituting (3.24) in (3.23), finally obtain

$$(C/N)_D = C/N \cdot \Delta N_U \qquad (3.25)$$

Applying decibels in expression (3.25), it is possible to write

$$(C/N)_D(dB) = C/N(dB) + \Delta N_U(dB) \qquad (3.26)$$

Example 3.7

The C/N values for a satellite link are 30 dB for the uplink and 14 dB for the downlink. Calculate the ΔN_U value.

Solution. Using (3.24), obtain

$$\Delta N_U = 10 \log \left(1 + \frac{10^{14/10}}{10^{30/10}} \right) = 0.1077 \, \text{dB}$$

3.5 Uplink Budget Analysis

Figure 3.5 shows the uplink thermal noise model.

Using (3.7) it is possible to write, for the uplink case

$$(C/N)_U = \frac{(\text{EIRP})_T (G/T)_S}{k L_a L_b B_T} \tag{3.27}$$

where $(\text{EIRP})_T$ represents the EIRP of the transmitting ground terminal, $(G/T)_S$ represents the G/T of the satellite's receiving section, and B_T is the transponder bandwidth.

The value of $(\text{EIRP})_T$ must warrant that the transponder operates in saturation, which is usual in broadcast applications. Therefore, the *power flux density* (PFD) at the input of the transponder must be equal to the SFD.

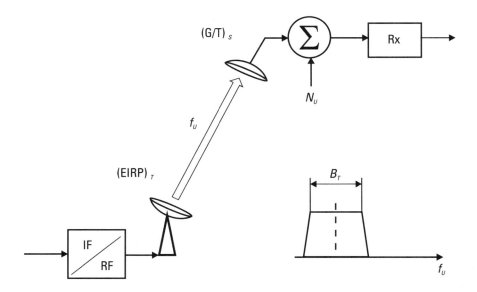

Figure 3.5 Uplink thermal noise model.

Then,

$$SFD = \frac{(EIRP)_T}{4\pi d^2}, W/m^2 \tag{3.28}$$

For the uplink case

$$L_b = \left(\frac{4\pi d}{\lambda_U}\right)^2 \tag{3.29}$$

and

$$\lambda_U(m) = \frac{0.3}{f_U(GHz)} \tag{3.30}$$

Substituting (3.28)–(3.30) into (3.27) and using decibels, it is possible to obtain

$$(C/N)_U(dB) = SFD(dBW/m^2) + (G/T)_S(dB/K) \\ - 20\log f_U(GHz) - B_T(dB.Hz) - L_a(dB) + 207.15 \tag{3.31}$$

Example 3.8

Determine the uplink C/N for a transmitting ground terminal sending a modulated carrier to Transponder 23, Galaxy V (129° W), (6.385 GHz). The ground terminal is located at 22° N, 80° W. The additional loss is 0.5 dB. The transponder operates in saturation and has a 36-MHz bandwidth. The SFD is –83.5 dBW/m² and (G/T)$_S$ is –1.6 dB/K.

Solution.   Using (3.31),

$$(C/N)_U = -83.5 - 1.6 - 20\log(6385) \\ - 10\log(36 \cdot 10^6) - 0.5 + 207.15 = 30 \text{ dB}$$

3.6 Downlink Budget Analysis

Figure 3.6 shows the downlink thermal noise model. In the analysis that follows, it is assumed that the TWTA (and the associated transponder) is

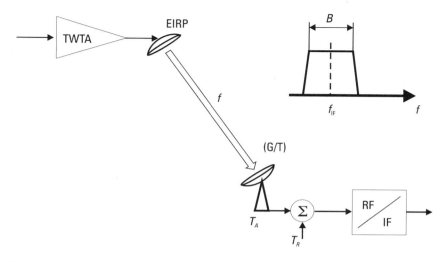

Figure 3.6 Downlink thermal noise model.

operating at saturation and in a single-carrier mode, which is typical in broadcast applications. Also it is assumed that there is not any interference at the input of the ground reception system. The frequency represents the downlink frequency band. For C-band systems this value can be considered to be 4 GHz, and for BSS Ku-band systems the typical value is approximately 12 GHz.

Again, using (3.7), the C/N for the downlink case can be expressed as

$$(C/N)_D = \frac{EIRP \cdot G/T}{kL_b L_a B} \qquad (3.32)$$

where EIRP corresponds to satellite power output, G/T corresponds to the receiving ground terminal, and B is the noise bandwidth of the IF amplifiers of the receiving ground terminal (also called the tuner). Because of the high selectivity of the IF amplifiers, the noise bandwidth and the signal bandwidth have almost the same numerical value. It is a common practice to call it the receiver or IRD bandwidth.

Substituting (3.25) in (3.32) and using decibels

$$\begin{aligned}(C/N)(dB) = \ &EIRP(dBW) + (G/T)(dB/K) - L_b(dB) \\ &- L_a(dB) - B(dB.Hz) - \Delta N_U(dB) + 228.6(dBW/K.Hz)\end{aligned} \qquad (3.33)$$

Example 3.9

Calculate the overall C/N for a satellite link with the following parameters:

- Satellite's EIRP: 45 dBW;
- Downlink free-space loss: 205.4 dB (12 GHz);
- Uplink noise contribution: 0.5 dB;
- Additional losses: 0.5 dB;
- Receiver ground terminal bandwidth: 20 MHz;
- Receiver ground terminal (G/T): 20 dB/K.

Solution.   The receiver ground terminal bandwidth, expressed in decibels, is

$$B(\text{dB.Hz}) = 10\log\left(20\cdot10^{6}\right) = 73\,\text{dB.Hz}$$

Using (3.33),

$$\text{C}\,/\,\text{N} = 45 + 20 - 205.4 - 0.5 - 73 - 0.5 + 228.6 = 14.2\,\text{dB}$$

3.7   Downlink Performance Analysis: C Band Versus Ku Band

Figure 3.7 shows the main characteristics of a satellite downlink for typical TV broadcast applications, for which the satellite transmitter antenna gain is fixed by the beamwidth implications of the coverage area requirement, and the receiver antenna size is as large as possible, considering convenience and cost. If A is the coverage area, the solid angle Ω is

$$\Omega = \frac{A}{d^{2}}\text{, steradians} \tag{3.34}$$

If it is assumed that the energy is concentrated in the main beam, the antenna gain G_T of the onboard transmitting antenna is inversely proportional to the beam's solid angle,

$$G_{T} = \frac{K_{p}}{\Omega} = \frac{K_{p}d^{2}}{A} \tag{3.35}$$

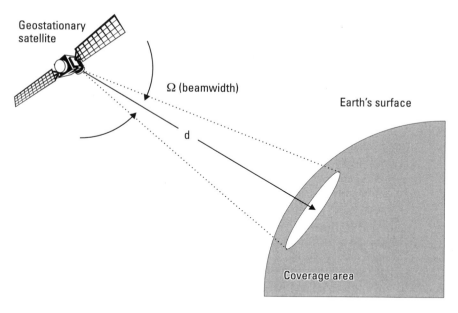

Figure 3.7 Satellite downlink for broadcast applications: fixed antenna gain at transmitter end and fixed antenna size at receiver end.

where K_p is the constant of proportionality. If the receiving antenna on the ground terminal has a fixed size, then the antenna gain G_R is

$$G_R = \eta \left(\frac{\pi D}{\lambda} \right)^2 \tag{3.36}$$

where η is the antenna's efficiency. Substituting (3.3), (3.35), and (3.36) into (3.1),

$$\mathrm{C/N} = \frac{P_T \left(\dfrac{K_p d^2}{A} \right) \cdot \left[\eta \left(\dfrac{\pi D}{\lambda} \right)^2 \right]}{k \left(\dfrac{4\pi d}{\lambda} \right)^2 \cdot L_a \cdot B} \tag{3.37}$$

Simplifying (3.37) obtains

$$C/N = \left[\frac{\eta K_p}{16\,Ak}\right] \cdot \frac{P_T D^2}{L_a BT} \tag{3.38}$$

where the factor in brackets has a constant value.

Now, it is possible to derive the following conclusions from (3.38):

- The C/N, as a performance measure of the satellite downlink, is no longer dependent on the carrier frequency. The reduced size of the antenna diameter on Ku-band ground-receiving terminals is because of the larger satellite EIRP used in these systems, compared with the lower-EIRP power-constrained C-band systems—not because the higher frequency carrier allows a higher value in receiving antenna gain. In both cases the satellite EIRPs are limited by regulatory PFD limits on the Earth's surface.

- The downlink performance is more sensitive to antenna diameter than the system's noise temperature.

- The additional loss L_a is higher in Ku-band systems (rain attenuation) than in C-band systems.

- The downlink performance is no longer dependent on distance. Nonetheless, the use of geostationary satellites involves a more viable technology and has a beneficial and economic impact on ground receiving terminals compared to any other kind of satellite (LEO and HEO).

Example 3.10

Determine the relationship between the antenna diameters of two ground receiving terminals, one operating in the C band (4 GHz) and the other in the Ku band (12 GHz). The ground terminal has the following characteristics:

- C-band and Ku-band transponders have the same EIRP value and bandwidth.

- The additional loss is 0 dB in the C band and 4 dB in the Ku band (rain attenuation).

- The system noise temperature in the C band is 55K, while in the Ku band it is 110K.

Solution. Let D_C and D_{Ku} represent the antenna diameters for the C band and the Ku band, respectively. Using (3.38) it is possible to write

$$\frac{(C/N)_C}{(C/N)_{Ku}} = \left(\frac{D_C}{D_{Ku}}\right)^2 \cdot 10^{0.4} \cdot \frac{110}{55} = 1$$

and

$$D_C = 0.45\, D_{Ku}$$

Then the antenna diameter could be lower in C-band systems than in Ku-band systems if both satellites had the same power output. However, note that in this scenario the C-band antenna is bigger by f^2 to yield the same gain pattern as that of the Ku-band spacecraft.

3.8 Radio Propagation Impairments in Satellite TV Links

The transmission of a satellite signal occurs almost in free-space conditions (more than 97% of the slant path). However, the troposphere (less than 100 km above the Earth's surface) and ionosphere (extending from 90 to 1,000 km above the Earth's surface) introduce significant impairments, whose importance depends on carrier frequency, elevation angle, atmosphere and ionosphere status, and solar activity. Rain influence is perhaps the most important single phenomenon over 10 GHz.

3.8.1 Gaseous Atmospheric Absorption

Absorption due to oxygen, water vapor, and other atmospheric gases, as distinguished from rain and other "hydrometeors," is basic and unavoidable. The attenuation is negligible at frequencies less than 10 GHz. There are specific frequency bands where absorption is high. The first band, caused by water vapor, is about 22 GHz, while the second band, caused by oxygen, is about 60 GHz. Absorption increases as the elevation angle is reduced. A variation given by cosec (EL°) can be applied to transform the attenuation over a zenith path. In the frequency range of 1 to 20 GHz, the zenith one-way absorption is approximately in the range 0.03 to 0.2 dB and, for 10° of elevation, the gaseous absorption is in the range 0.17 to 1.15 dB.

3.8.2 Ionospheric Scintillation

Scintillation is a rapid fluctuation of signal amplitude, phase, polarization, or angle of arrival. In the ionosphere, scintillation occurs because of multipath due to small variations of the refractive index caused by local concentrations of ions. Ionospheric scintillation decreases as $1/f^2$ and has a significant effect below 4 GHz. Tropospheric scintillation can also occur, again with multipath as a cause, and in this instance increases with frequency. Generally this is small or negligible, amounting to tenths of decibels at the Ku band.

3.8.3 Faraday Rotation

The electrons in the ionosphere along with the Earth's magnetic field cause a rotation in the plane polarization known as the Faraday rotation. The rotation diminishes inversely with the square of the frequency and sometimes has a maximum value of 150° at 1 GHz. At 4 GHz, the Faraday rotation is 9°, and it can be neglected above 10 GHz.

3.8.4 Rain Influence

The main effect of rain above 10 GHz is to attenuate the signal and, because of its behavior as a lossy attenuator, to increase the antenna noise temperature. In addition to these effects, rain has a depolarizing effect, which creates a cross-polarized component with linear polarization and a loss of circular polarization. The attenuation is caused by the scattering and absorption of radioelectrical waves by drops of liquid water. The attenuation increases as the wavelength approaches the size of a typical raindrop, which is about 1.5 mm.

3.8.4.1 Rain Attenuation

The calculation of rain attenuation can be divided into two main steps. The first one is to estimate the rain rate R in mm/hr as a function of the cumulative probability of occurrence probability $(r \leq R)$, where r is the random variable rain rate and R is a specific value. This probability helps to determine the grade of service to be provided and thus the values of margin required. The second step is to calculate the attenuation resulting from those rain rates, given the elevation angles, the ground terminal latitude, and the carrier frequency.

There are two authoritative rain models that are widely used: Crane Global [1] and ITU-R [2]. The Crane Global model is an empirically based model that uses data from geographical regions to develop a relationship between the path average rain and the point rain rate. A revision of this model that accounts for both the dense center and fringe area of a rain cell is the so-called two-components model. The ITU-R model is the empirically based model recommended by the ITU. The model calculates the attenuation due to a rain rate that occurs 0.01% of the time. Then the model uses a reduction factor and an interpolation procedure to determine the rest of the distribution. The model is based on point rainfall statistics.

Both models use the following basic expressions to predict rain attenuation:

$$\gamma_R = kR^a \tag{3.39}$$

$$A_R = \gamma_R \cdot L_e \tag{3.40}$$

where γ_R is the specific attenuation (in decibels per kilometer) and R is the point rain rate (in millimeters per hour) for a specific outage. The k and α values are calculated from theoretical formulae and L_e represents the effective path length (in kilometers) through rain. The rain rate is a measure of the average size of the raindrops. When the rain rate increases (it rains harder), the raindrops are larger and thus there is more attenuation. Reference [3] reports a method that combines rain attenuation and other propagation impairments along Earth satellite paths. For system planning, the method mentioned in Appendix B is good enough for engineering practice. At the C band, the rain attenuation has practically a negligible effect. At the Ku band, the attenuation ranges between 2 and 10 dB and, although it is a large value, is manageable in the link budget. However, at the downlink frequency of 20 GHz, the attenuation for equivalent link availability would be higher than 10 dB.

3.8.4.2 Antenna Noise Temperature Increase Due to Rain

In addition to causing attenuation, rain increases the downlink system noise temperature. The physical reason is that rain acts like an attenuator and that any warm attenuator produces additional thermal noise. The thermal noise rain model is illustrated in Figure 3.8.

The antenna noise temperature under rain conditions, T_A', is

$$T_A' = 10^{-A_R/10} \cdot T_{sky} + \left(1 + 10^{-A_R/10}\right) \cdot T_0, \text{K} \tag{3.41}$$

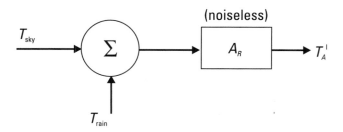

Figure 3.8 Thermal noise rain model. T_{sky} is the sky noise temperature, T_{rain} is the rain noise temperature, and T_A' is antenna noise temperature under rain conditions. A_R is the attenuation in decibels.

where T_0 is the thermodynamic rain temperature. The antenna noise temperature increment ΔT_A can be calculated as

$$\Delta T_A = 10^{-A_R/10} \cdot T_{sky} + \left(1 - 10^{-A_R/10}\right) \cdot T_0 - T_{sky}, \mathrm{K} \qquad (3.42)$$

Rearranging (3.42),

$$\Delta T_A = \left(1 - 10^{-A_R/10}\right) \cdot \left(T_0 - T_{sky}\right) \qquad (3.43)$$

The difference $\left(T_0 - T_{sky}\right)$ can be approximately substituted by 240K [4] in the 12-GHz band (downlink). Then,

$$\Delta T_A = 240 \cdot \left(1 - 10^{-A_R/10}\right), \mathrm{K} \qquad (3.44)$$

3.8.4.3 Depolarization

Rain also changes the polarization of the electromagnetic wave. Due to resistance of the air, a falling raindrop assumes the shape of an oblate spheroid. Consequently, the transmission path length through the raindrop is different for different carrier polarizations, and the polarization of the received carrier is altered. For a satellite communication system with dual linear polarizations, the change in polarization has two effects. First, there is a loss in the carrier strength because of misalignment of the antenna relative to the clear sky orientation. Typical values are less than 0.5 dB. Second (and usually more significant), there is additional interference due to the admission of a component of any cofrequency carrier in the opposite polarization.

3.8.5 Statistical Analysis of Rain Fading

In the design of any engineering system, it is impossible to guarantee the performance under each possible condition. One sets reasonable limits based on the conditions that are likely to occur at a given level of probability. Also, in the design of a satellite link, a power margin is included to compensate the effects of rain at a given level of service interruption (outage). The rain fading can be analyzed using a flat fading model, because a well-designed, fixed ground terminal is not affected with shadowing, blockage, or delayed multipath rays, and rain scattering is almost negligible [5]. Then, the analysis can be mathematically treated on the basis of the received carrier power in terms of C/N at the demodulator input, as is shown in Figure 3.9.

The margin M_0 relative to the required value $(C/N)_0$ to achieve the desired quality performance is defined as [6]

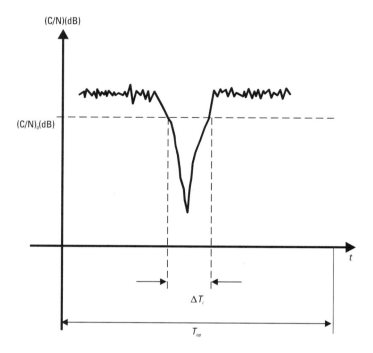

Figure 3.9 C/N temporal variations due to rain fading; T_{op} is a reference time.

$$M_0 = \frac{C/N}{(C/N)_0} \geq 1 \qquad (3.45)$$

Using decibels,

$$M_0(\text{dB}) = C/N(\text{dB}) - (C/N)_0(\text{dB}) \geq 0 \qquad (3.46)$$

The probability of outage is the percentage of time at which the system performance is unacceptable and it is represented as p (%). It can be defined as

$$p(\%) = \text{Prob}\left(M_0 < 0\,\text{dB}\right) \cdot 100 = \frac{\displaystyle\sum_{\forall i} \Delta t_i}{T_{op}} \cdot 100 \qquad (3.47)$$

Example 3.11

The system's availability is the complementary event of the probability of outage. If the system's availability is 99%, determine the time outage in a year (average).

Solution. The time outage in a year (average) is

$$\frac{1}{100} \cdot 365 = 3.65 = 3 \text{ days, } 15 \text{ hours, } 36 \text{ minutes}$$

During this time the system cannot fulfill the expected quality performance, averaged in a year.

Let us consider now the additional loss L_a. It may be split as

$$L_a = L_R + \Sigma L, \text{dB} \qquad (3.48)$$

where L_R represents, in decibels, the rain loss as a stationary random process and ΣL (also in decibels) represents other time-invariant losses.

The probability of outage can also be written as

$$p(\%) = \text{Prob}\left(L_R \geq A_R, \text{dB}\right) \cdot 100 \qquad (3.49)$$

where A_R is a specific numerical value of the random variable L_R. Since one knows the cumulative probability distribution law in (3.49), it is possible to solve it for A_R. The value thus obtained is represented by $A_R(p\%)$ to specify the selected value of probability outage. Substituting (3.33) in (3.46) and using the developed concepts through (3.48) and (3.49), it is possible to write

$$M_0 + A_R = \text{EIRP} + G / T - L_b - \Sigma L - B \\ -\Delta N_U - (C / N)_0 - 228.6, \text{dB} \tag{3.50}$$

Let $(G/T)'$ be the receiving ground terminal figure of merit in rain conditions and G/T be the same in clear-sky conditions. Taking into account the definition of G/T (3.6), then

$$\frac{(G / T)'}{G / T} = \frac{G_R / (T_A + \Delta T_A + T_R)}{G_R / (T_A + T_R)} = \frac{1}{1 + \dfrac{\Delta T_A}{T_A + T_R}} \tag{3.51}$$

where T_A is the antenna noise temperature in clear-sky conditions and T_R is the receiver noise temperature. If one defines a factor ΔT (not to be confused with a real temperature increase), in decibels, as

$$\Delta T(\text{dB}) = 10 \log\left(1 + \frac{\Delta T_A}{T_A + T_R}\right) \tag{3.52}$$

then

$$(G / T)'(\text{dB}) = G / T(\text{dB}) - \Delta T(\text{dB}) \tag{3.53}$$

Substituting (3.44) into (3.52) and using a maximum value of 50K in T_A (typical in Ku-band offset antennas),

$$\Delta T = 10 \log\left[1 + \frac{240\left(1 - 10^{-A_R/10}\right)}{50 + 290\left(10^{F_R/10} - 1\right)}\right] \tag{3.54}$$

where F_R is the receiver noise figure in decibels and is another way to characterize the receiver noise, as is the noise temperature T_R. The relationship between F_R (adimensional) and T_R is the well-known formula

$$F_R = 1 + \frac{T_R}{290} \tag{3.55}$$

where T_R is in Kelvin. Note that in (3.55), F_R is linear.

Taking into consideration the antenna noise increment due to rain, (3.50) can be rewritten as

$$\begin{aligned} M_0 + A_R &= \text{EIRP} + (\text{G} / \text{T})' - L_b - \Sigma L \\ &- B - \Delta N_U - (\text{C} / \text{N})_0 + 228.6 \end{aligned} \tag{3.56}$$

Substituting (3.53) into (3.56) and rearranging some terms,

$$\begin{aligned} M_R(\text{dB}) &= \text{EIRP}(\text{dBW}) + \text{G} / \text{T}(\text{dB} / \text{K}) - L_b(\text{dB}) \\ &- \Sigma L(\text{dB}) - B(\text{dB.Hz}) - \Delta N_U(\text{dB}) - (\text{C} / \text{N})_0(\text{dB}) + 228.6 \end{aligned} \tag{3.57}$$

where

$$M_R(\text{dB}) \geq A_R(p\%) + \Delta T \tag{3.58}$$

3.9 Application to Satellite TV System Design: Downlink Margin Equation

Equations (3.57) and (3.58) constitute the formal basis to the downlink budget analysis and dimensioning of satellite TV systems. They can be used in a straightforward way for Ku-band analog systems. For Ku-band digital systems, they can be further developed, remembering that [7]:

$$\begin{aligned} (\text{C} / \text{N})_0(\text{dB}) &= (E_b / N_0)_0(\text{dB}) \\ &+ R_b(\text{dB.bps}) - B(\text{dB.Hz}) - G_C(\text{dB}) \end{aligned} \tag{3.59}$$

where $(E_b/N_0)_0$ is the average bit energy-to-noise spectral density ratio for a prescribed BER; and a specific kind of modulation, R_b, denotes the MPEG-2 transport stream information bit rate (see Chapter 4); and G_C is the coding gain of the channel encoder-decoder (see Appendix C). Substituting (3.59) into (3.57), one obtains

$$M_R(\text{dB}) = \text{EIRP}(\text{dBW}) + \text{G}/\text{T}(\text{dB}/\text{K}) - L_b(\text{dB}) - \Sigma L(\text{dB})$$
$$-R_b(\text{dB.bps}) - \Delta N_U(\text{dB}) + G_C(\text{dB}) - (E_b/N_0)_0(\text{dB}) + 228.6 \tag{3.60}$$

which can be applied with (3.58) in associated digital cases.

Example 3.12

Determine the rain margin for the downlink design in a satellite TV system operating in the Ku band (12 GHz) with H polarization. The ground terminal is located in 22° N, 80° W and uses a receiver with a 0.8-dB noise figure; in addition, its height above mean sea level is 200m, and the elevation angle pointed to the satellite is 30°. Consider 99% of availability (average year).

Solution.   According to the map of Figure B.1 (Appendix B), the ground terminal is in zone N, and $R_{0.01}$ is 95 mm/hr (Table B.1). The corresponding values of k_H, k_V, α_H, and α_V are 0.0188, 0.00168, 1.217, and 1.2, respectively (Table B.2). The value of $\tau = 0°$, and k and α values are

$$k = \frac{[0.0188 + 0.0168 + (0.0188 - 0.0168) \cdot \cos^2 30°]}{2} = 0.01855$$

$$\alpha = \frac{[(0.0188)(1.217) + (0.0168)(1.2) + (0.023 - 0.020)\cos^2 30°]}{(2)(0.01855)} = 1.214$$

The specific rain attenuation is

$$\gamma_{R0.01} = 0.01855(95)^{1.214} = 4.67 \text{ dB/km}$$

Other parameters' values are

$$L_S = \frac{5 - 0.2}{\sin 30°} = 9.6 \text{ km}$$

$$L_O = 35.e^{-0.015(95)} = 8.42 \text{ km}$$

$$r_{0.01} = \cfrac{1}{1 + \cfrac{9.6}{8.42} \cdot \cos 30°} = 0.503$$

The effective length of the rainy path is

$$L_e = 9.6 \cdot 0.503 = 4.83 \text{ km}$$

Then, the rain attenuation for $p = 0.01\%$ is

$$A_{R_{0.01}} = 4.67 \cdot 4.83 = 22.5 \text{ dB}$$

The rain attenuation for $p = 1\%$ can be calculated as

$$A_R(1\%) = 22.5 \cdot 0.12 \cdot (1)^{-(0.546 + 0.043 \cdot \log 1)} = 2.7 \text{ dB}$$

Using (3.54), one obtains

$$\Delta T = 10 \log \left[1 + \frac{240 \cdot \left(1 - 10^{-0.27} \right)}{50 + 290 \cdot \left(10^{0.08} - 1 \right)} \right] = 3.06 \text{ dB}$$

The rain margin, using (3.58), is

$$M_R \geq 2.7 + 3.06 = 5.76 \text{ dB}$$

which can be rounded up to 6 dB.

Example 3.13
Repeat Example 3.12 using a 99% worst-month availability.

Solution. The average year outage is (see Appendix B)

$$p = 0.3 \cdot (1)^{1.15} = 0.3\%$$

The rain attenuation is now

$$A_R(\%) = 22.5 \cdot 0.12 \cdot (0.3)^{-(0.546 + 0.043 \cdot \log 0.3)} = 5.07 \text{ dB}$$

The value of ΔT is

$$\Delta T = 10 \log\left[1 + \frac{240\left(1 - 10^{-0.507}\right)}{50 + 290\left(10^{0.08} - 1\right)} \right] = 4.017 \text{ dB}$$

The rain margin is

$$M_R \geq 5.07 + 4.017 = 9.087 \text{ dB}$$

and, finally, $M_R = 10$ dB.

References

[1]   Crane, R. K., "Prediction of Attenuation by Rain," *IEEE Trans. on Comm.,* Vol. COM-28, No. 9, September 1980, pp. 1717–1733.

[2]   ITU-R, *Rec. ITU-R PN.837-1, Rec. 838, Rec. 839,* 1992–1994.

[3]   Dissanayake, A., J. Alnutt, and F. Haidara, "A Prediction Model That Combines Rain Attenuation and Other Propagation Impairments Along Earth-Satellite Paths," *IEEE Trans. on Antennas and Propagation,* Vol. 45, No. 10, October 1997, pp. 1546–1558.

[4]   Morgan, W. L., and G. D. Gordon, *Communications Satellite Handbook,* New York: John Wiley and Sons, 1989.

[5]   Matricciani, E., and C. Riva, "Evaluation of the Feasibility of Satellite System Design in the 10–100 GHz Frequency Range," *International Journal of Satellite Communications,* Vol. 16, 1998, pp. 237–247.

[6]   Sklar, B., *Digital Communications: Fundamentals and Applications,* Englewood Cliffs, NJ: Prentice Hall, 1988, pp. 188–244.

[7]   Sklar, B., "Defining, Designing, and Evaluating Digital Communication Systems," *IEEE Communications Magazine,* November 1993, pp. 92–101.

4

Transmission Techniques and Standards for Satellite TV Signals

4.1 Analog Transmission Techniques

4.1.1 Baseband Audiovisual Analog Signals

The baseband audiovisual signals used in analog satellite TV systems are very nearly the same as those used in analog TV terrestrial systems—the NTSC, PAL, and SECAM standards. Figure 4.1 shows the spectral composition of these standards corresponding to satellite TV broadcast applications. It should be observed that the sound subcarrier frequency is different from conventional TV standards.

The NTSC, PAL, and SECAM standards have some inherent weaknesses, especially when they are used on satellite links. The main drawbacks are described as follows:

- Performance quality was partially sacrificed to achieve the goal of broadcasting signals that could be received by both color and black-and-white television receivers. The frequency multiplexing of the chrominance and luminance signals into the composite video signal creates an environment in which interactions between these components can occur; this results in picture distortions, known as cross color and cross luminance. These effects are more pronounced on

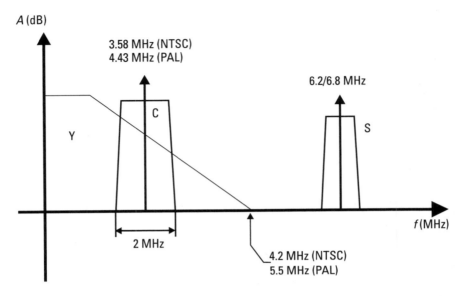

Figure 4.1 The spectrum of the composite television signal for satellite broadcasts (Y: luminance component; C: chrominance component; and S: sound component).

larger screen TV sets. Interactions between the audio and video signals may also result in distortions (sound bars and beat interference).

• With satellite transmission using *frequency modulation* (FM), the chrominance component is more affected than the luminance component (because of the quadratic-law output spectral noise density of FM detectors), and it results in color distortion.

In the early 1980s, broadcast engineers began to develop a new television standard known as the multiplexed analog component system. The components sound/data/sync, chrominance, and luminance were time-multiplexed (Figure 4.2).

In the MAC system, the whole bandwidth is exclusively available for its TDM, which eliminates chrominance interactions and can result in better quality reproduction. Color distortion is minimized, and the available color bandwidth is increased. Although ahead of its time when first introduced, it was a technology in the transition between analog and digital systems. The

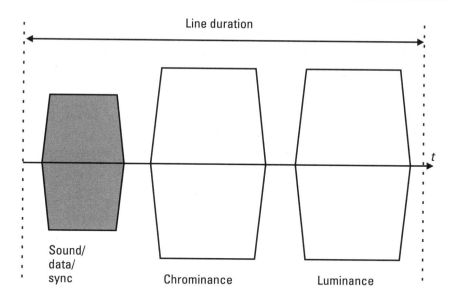

Figure 4.2 MAC format.

MAC system has been overtaken by technical progress in digital TV systems. Nevertheless, the MAC system is still in use [1].

4.1.2 Picture Quality

The *luminance-to-noise ratio* (S/N) is taken as the quality of analog television signals. The usual practice is to use an index Q of quality, which refers to the five-point scale appearing in the ITU-R Recommendation. This scale is shown in Table 4.1.

Table 4.1
Five-Point Scale for Evaluating Image Quality

Q (Quality)	Impairment
5 (Excellent)	Imperceptible
4 (Good)	Perceptible but not annoying
3 (Fair)	Slightly annoying
2 (Poor)	Annoying
1 (Bad)	Very annoying

It is possible to write an empirical relationship between S/N and Q when it is used in FM transmission on satellite links:

$$(S/N)_{uw} = 23 - Q + 1.1 \cdot Q^2, dB \qquad (4.1)$$

where $(S/N)_{uw}$ represents the unweighted luminance-to-noise ratio where the videometric and pre- and deemphasis effects are not taken into account. For instance, if one defined the specific index of quality, it is possible to know, using (4.1), the required S/N. Then:

$Q = 4.8$ contribution (studio quality); S/N = 43.5 dB;

$Q = 4.5$ primary distribution; S/N = 40.7 dB;

$Q = 4.2$ secondary distribution (broadcasting); S/N = 38.2 dB.

For planning purposes, the ITU-R has recommended a value of $Q = 3.5$, and it should be considered the worst quality of all satellite broadcast applications. Using (4.1), a value of 33 dB is found for the unweighted S/N, and it must be achieved at the following times, at the least:

- At the limit of the service area;
- During 99% of the least-favorable month (99.7% average year), taking only into consideration the radio propagation effects on signal degradation;
- At the end of the useful life of the satellite and ground receiver.

4.1.3 FM-TV

FM is only used for analog satellite broadcasting. As the modulated carrier envelope is constant (the carrier amplitude is not affected by the modulating signal), it is robust with respect to the nonregenerative satellite transponder. However, for a given quality of link, it offers the useful possibility of a trade-off between the C/N and the bandwidth occupied by the carrier. The audio-visual signal shown in Figure 4.1 is used as the modulating signal in the FM system.

The two fundamental equations for satellite FM-TV are described as follows:

- A relation that determines the transmission bandwidth as a function of the frequency deviation;
- The relation between S/N and C/N.

The equivalent transmission noise bandwidth B should have a bandwidth equal to the transponder bandwidth because of the high selectivity of the overall frequency response of the satellite channel (including the ground receiver frequency response). This noise bandwidth can be accurately calculated using Carlson's rule. That is,

$$B = \Delta f_{pp} + 2 f_{max} \qquad (4.2)$$

where f_{max} is the maximum frequency in the spectrum of the baseband video signal.

The second equation may be written in the well-known form

$$(S/N) = \frac{3}{2} \left(\frac{\Delta f_{pp}}{b_n} \right)^2 \cdot \left(\frac{B}{b_n} \right) \cdot pw \cdot C/N \qquad (4.3)$$

$$\text{if } C/N \geq \text{ threshold}$$

where:

S/N: S/N (weighted) after FM demodulation;

C/N: C/N at the FM demodulator input;

B: Equivalent transmission noise bandwidth;

b_n: Bandwidth of noise measured at the receiver output;

Δf_{pp}: Peak-to-peak frequency deviation produced by a 1-V video signal (standard 1-V level corresponds to the interval, in the absence of preemphasis or at the preemphasis transition frequency, between the bottom of the synchronizing pulses and the maximum luminance level);

pw: Represents the combined effect of pre- and deemphasis and video metric weighting. (Preemphasis is a technique used to boost high frequencies before transmission. Since noise density increases with frequency, the subsequent deemphasis reduces the signal level to a normal value and consequently attenuates the high frequency noise added during the transmission path. The videometric-weighted factor takes into account the nonflat curve of human eye sensitivity as a function of video frequency and reduces the baseband noise within the video signal bandwidth.) Table 4.2 gives the values of pw for various TV standards (ITU-R, Report 637).

Table 4.2
Improvement in S/N Provided by Pre- and Deemphasis and Videometric Weighting
(ITU-R, Report 637)

TV Standard (Lines per Frame/ Frames per Second)	TV Standard Name	bn (MHz)	pw (dB)
525/60	NTSC-M	4.2	12.8
	NTSC (Unified)	5.0	14.8
625/50	PAL (B,G,H)	5.0	16.3
	PAL (I)	5.0	12.9
	Unified	5.0	13.2
	PAL (D,K,L)	6.0	18.1

Example 4.1

Parameters for PAL (unified) TV signal transmission by ASTRA satellites are listed as follows:

Δf_{pp} = 13.5 MHz/V;

f_{max} = 6 MHz;

bn = 5 MHz;

pw = 13.5 dB.

Find a C/N value for S/N = 45 dB and the required transponder bandwidth.

Solution.   Using (4.2), obtain

$$B = 13.5 + 2 \cdot 6 = 25.5 \, \text{MHz}$$

Then, using (4.3),

$$(C / N) = 45 - 10 \log\left(\frac{3}{2} \cdot \left(\frac{13.5}{5}\right)^2 \cdot \left(\frac{25.5}{3}\right) \right) - 13.2 = 13 \, \text{dB}$$

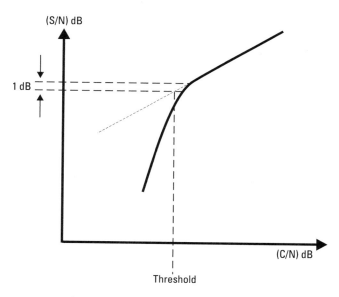

Figure 4.3 Relation between S/N and C/N in an FM demodulator used in satellite broadcasting.

Equation (4.3) is valid above a certain level known as the *threshold* (Figure 4.3), where S/N is proportional to C/N. Below this threshold very complex phenomena cause the S/N to deteriorate very rapidly.

Interference must also be included in the overall C/N as *cochannel interference* (CCI) and *adjacent channel interference* (ACI). Recommended values are shown in WARC'77 and RARC'83 for intended analog systems. These values deal with signal impairments on the video information (see Table 1.3).

The threshold condition is seen on a TV picture by the appearance of flashing. It can be experimentally shown that the threshold lies in the region of a C/N ratio of 8 dB; below this value the picture rapidly becomes very noisy. It is necessary, therefore, to operate FM systems above this threshold, but in the particular case of satellite broadcasting, it must work as closely as possible to the threshold to economize the radiated power.

There are currently three basic types of extended threshold demodulation techniques: the extended threshold PLL demodulator, the *dynamic tracking filter* (DTF) demodulator, and the *frequency feedback loop* (FFL) demodulator. All of these techniques are based on the principle of reducing the detection bandwidth of the demodulator, leading to a significant reduction in noise power while leaving the signal power relatively unchanged. A basic understanding of these techniques can be found in [2]. Typical threshold figures using the above-mentioned techniques are in the 5- or 6-dB range. In practice, it is advisable to include a small safety margin in relation to the real threshold and to work with a reference threshold.

The preceding discussion applies only to the visual signal (monochrome or color), but the transmission of a television program demands at least one sound component, in particular the use of a narrowband FM-modulated subcarrier placed above the baseband video, time multiplexing within the field-blanking interval (sound-in-sync), or even the use of a special FM-modulated carrier transmitted in another transponder. In addition to the primary audio subcarrier, this principle can be extended to cover the transmission of several sound channels (a stereo sound signal and more multilingual audio channels). Some low-cost receivers are totally monophonic and tend to rely on the primary audio subcarrier demodulation with the aid of an intercarrier circuit, and this is the simplest design. However, if the receiver is designed to demodulate narrowband subcarriers for sound-broadcasting purposes, it is necessary to add more complexity to the receiver. Using several separate subcarriers introduces intermodulation products, so it is necessary to achieve a high degree of stability in the local-oscillator frequency (because of the relatively narrowband) or it would be incompatible with inexpensive receivers. A more detailed discussion of analog audio modulated subcarriers is found in [3].

4.2 Digital Transmission Techniques

4.2.1 Introduction

At WARC'77, a part of the Ku band (10.75–12.75 GHz) was assigned for satellite TV broadcasting. FM-modulated PAL, SECAM, and NTSC could be used with channels of, for instance, 27 MHz. A C/N of about 14 dB is needed to provide good reception of these analog TV signals.

Improvements in satellite and receiver technology have made it possible to receive analog TV signals transmitted on medium-powered European Ku-band satellites (such as ASTRA) with a 60-cm diameter parabolic antenna. Under these conditions, the step toward DBS systems was easily taken. Analog Ku-band DBS systems (also known in Europe as DTH services) have expanded significantly in Europe.

In other parts of the world, for example, North and South America and the Far East, satellites broadcasting TV generally use the C band (3.7–4.2

GHz in the downlink) and analog transmission. The choice of the lower frequency C band instead of the Ku band was originally motivated by technology (20 years ago C-band hardware was dramatically less expensive than that for the Ku band) and better propagation through rain, especially in the tropics. The receive antenna diameter ranges between 2 and 4m, making C-band satellite TV in urban areas relatively less attractive in this frequency band now that Ku-band electronics bear a negligible price premium.

Digital satellite transmission has some clear advantages, as compared with analog satellite transmission. These advantages are described as follows:

- Significant improvement in digital source coding: New low complexity algorithms have become available for the compression of video signals that have been standardized by the Motion Picture Expert Group (MPEG) in the MPEG-2 standard. Video compression now makes it possible to transmit a single video program in not more than 10–20% of the bandwidth needed for analog transmission, which saves space segment costs.

- Smaller required C/N: Compared to analog video signals, which typically require a C/N of about 10–14 dB, digital can work with C/N values of 4–8 dB, which reduces the amount of power needed to transmit an acceptable signal.

- Flexible multiplexing that allows for any combination of multiple video, audio, text, and multimedia streams to be put together into one single aggregated stream (known as a "multiplex") per transponder. This multiprogramming allows efficient bandwidth allocation between the different services on the transponder.

The first advantage especially triggered the fast introduction of digital satellite broadcasting. Although analog DBS was growing in most parts of the world, the largest use of all analog satellite transmission was for primary distribution from a TV studio to other TV studios or to CATV head ends (Figure 4.4). All these links were characterized by a limited number of receivers per channel. However, the usage fee for a satellite transponder needed to transmit a single (analog) TV program is still significant. By digitizing these distribution links, the program providers can save enormous amounts of money, thus justifying the investments in digital uplink and receive equipment. Processing at studios themselves has been digital for some time, so to transmit digital also represents a relatively natural technical evolution.

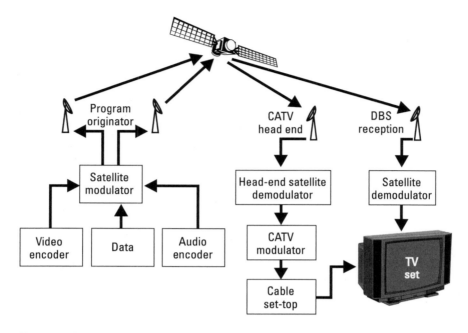

Figure 4.4 Typical digital satellite link.

4.2.2   The European Satellite Standard

In Europe, the first projects in the field of digital television broadcasting were directed toward terrestrial transmission. In different regions, many separate projects existed mostly without any mutual links. To serve as an umbrella for these projects, and to attract the attention of European politicians, the *European Launching Group* (ELG) was formed in 1992 with participants from most European public broadcasters, network providers, and TV industries. Very soon awareness grew in the ELG to quickly start digital services.

In September 1993, the ELG framed a memorandum of understanding (MoU) and transformed itself into the European Project on DVB (EP-DVB). One of the main points in the MoU was the adoption of the MPEG-2 standard for video coding and multiplexing. Within a few months, the EP-DVB approved the draft specifications for digital broadcasting by satellite (also called the DVB-S standard) and sent it to ETSI for further evaluation. The result of this work is the *European Telecommunication Standard* (ETS) ETS 300 421, which describes the framing structure, channel coding, and modulation for satellite transmission. This standard has been submitted to the ITU for consideration as a world standard. At its meeting in

Melbourne, Australia, in June 1995, the global prestandardization organization *Digital Audio Video Interactive Council* (DAVIC) accepted the ETS 300 421 system as the unique mechanism for digital satellite broadcasting in the Ku band.

Although the standard ETS 300 421 fulfills the requirements established in WARC'77, it is not tied to these parameters and can be used with other communication satellite constraints, for example, other transponder bandwidths different than 27 MHz.

4.2.3   Situation in the United States

The United States' several large systems began digital broadcasting before the European DVB standardization activity was complete. Consequently, the hardware used was generally not DVB-compatible, and the digital transmission systems market was considerably fragmented. Some representative systems are described as follows.

- For satellite delivery to head ends, the Digicipher system by *General Instruments* (GI) was proposed, although the extent of its success in actual commercial deployment is not clear. The Digicipher system uses proprietary algorithms for source coding, channel coding, and transport mechanisms.

- For DBS satellite transmission in the United States, GM/Hughes and Thomson developed DirecTV with *US Satellite Broadcasting* (USSB) as a partner. Four Ku-band satellites have already been launched using 32 transponders with more than 150 NTSC compressed digital programs following the recommendations of RARC'83. The modulation scheme is *quadrature phase shift keying* (QPSK) with concatenated convolutional and RS *forward error correction* (FEC). The source coding and transport scheme uses a combination of algorithms described within the MPEG-1 and MPEG-2 standards and can be suitably designated MPEG-1+. The DSS receiver system became available as a consumer product with at least 1 million units sold in the first year of operation in 1994. There are currently around 8 million subscribers with DSS systems.

- Another service provider is EchoStar, which uses fully DVB/MPEG-2–compliant technology. There are currently around 3 million EchoStar subscribers.

Recently, the ATSC Standard [4], which is very similar to the ETS 300 421 standard, was adopted.

4.3 MPEG-2 Overview

Before the affordable implementation of compression methods in the late 1980s, digital television had a serious limitation: It required an enormous transmission bandwidth. For example, the ITU-R standard 601 needs 216 Mbps.

Compression methods are used to reduce the information bit rate in a digital video signal to a fraction of its original value by removing redundancy and irrelevance. Redundancy relates to statistical properties of the image. The level of a signal at any moment can, to a certain extent, be predictable from its value in the past because picture sample values are related to each other, not only in the same line but also in the previous frames. Irrelevance relates to the physiological limitations of the *human visual system* (HVS) viewing an image. Depending on the picture content, the human eye can tolerate a certain amount of distortion leading to a lossy compression method. (Compression coding can be lossless or lossy. The first choice necessitates that, in the absence of channel errors, the signal at the transmission end and the decoding at the receiving end be identical.) In video applications, strict accuracy of information transmission is considered not vital, so lossy compression is considered essential for achieving commercial feasibility. Lossy compression has the significant advantage of reducing the bit rate compared to the corresponding lossless bit rate by (typically) at least an order of magnitude in bit rate. The optimum tradeoff requires subjective testing of the viewer's tolerance to image distortion. Basics on video-compression methods are reported in [5, 6].

When compression technology had advanced to the point where high-volume applications were possible, a clear standard was needed to ensure compatibility between systems from various manufacturers. This would help achieve economies of scale and thus lower prices.

In the late 1980s, the *International Standards Organization* (ISO) [7–9] began the process of drafting a series of worldwide standards that would underlie the use of digital compression in both videoconferencing and digital video applications. The *Joint Photographic Expert Group* (JPEG) led the effort for still photos and graphics, and the MPEG was the developer of motion picture applications.

The MPEG group had three major goals:

- To produce flexible and generic world standards for video and audio coding;

- To define features suitable for use in transmission and recording media;

- To create procedures used to evaluate systems and define a bit stream structure to enable coder and decoder designs.

MPEG-2, completed in 1994, set the stage for rapid growth in video-conferencing applications, the oldest and most evolved use of compressed digital video. It originally defined the system to provide VHS-quality pictures at bit rates up to 1.5 Mbps.

Created to be reverse-compatible to MPEG-1, MPEG-2 implements a flexible architecture to allow a multistandard environment where one system is able to output video and audio signals in *low-definition TV* (LDTV); *standard-definition TV* (SDTV) like NTSC, PAL, and SECAM; and HDTV. The most important contribution of MPEG-2 is that it gives an integrated transport mechanism for multiplexing the video, audio, and other data through packet generation and TDM. It is a layered approach in which the outer layers provide the system-level functions of synchronization and multiplexing, necessary for using one or more data streams within the same system; the inner layers define the image and audio-compression algorithms. The definition of the syntax (bit structure) includes a bit stream, a set of source-coding algorithms, and a multiplexing format to combine video, audio, and data. The system is scalable in that a variety of forms of lossless and lossy transmission are possible, along with the capability to uphold HDTV standards.

The source-coding algorithms are a mixture of three processes: (1) predictive coding, which also uses motion estimation and compensation to make use of temporal redundancy in the moving pictures; (2) transform coding, which uses *discrete cosine transform* (DCT) to make use of spatial redundancy; and (3) Huffman coding (run-length coding) to take out remanent redundancy from the bit stream produced by the first two processes. The source-coding algorithm also relies upon a buffer store that is used to regulate and smooth the flow of data and whose state of occupancy controls (by means of a feedback path) the coding precision: The more occupied the buffers are, the coarser the coding becomes to decrease the total number of data entering the buffer and vice versa.

To organize potential combinations enabled by the possible use of different source-coding techniques and the possible support of different coding parameter values as defined by the video standard, "profiles" and "levels" have been created as part of MPEG-2 video. Each profile [including the *signal-to-noise ratio* (SNR)] defines a different subset of compression tools, increasing in complexity and implementation cost, and is completely backward-compatible with previous profiles in the whole set. A level contains general parameters, such as image size, information bit rate, and decoder buffer size. Table 4.3 shows the various MPEG-2 coding levels and profiles. For SDTV broadcast applications, the relevant combination of profile and level is the so-called main profile at main level (MP@ML). With this MP@ML it is possible to transmit digital compressed TV programs with data range from 4 to 6 Mbps for standard picture quality that is subjectively equivalent to current analog NTSC, PAL, and SECAM standards. Compressed HDTV signals can be transmitted using an approximate information bit rate of 14 Mbps.

The MPEG-2 Layer 2 specification is used for audio coding and is based on *masking pattern–adapted universal subband integrated coding and multiplexing* (MUSICAM). The system provides near-CD quality at very low bit rates and is flexible in that mono, stereo, surround-sound, or multilingual audio can be transmitted [10]. This digital audio compression method takes advantage of the domination of one sound element against those in close proximity, but lower level background sounds or noise would not be heard even if reproduced with fidelity. The redundant information is not coded, so MPEG audio is a lossy compression scheme. Although MPEG-2 defines a complete protocol set for audio coding, other systems like Dolby digital AC3, which was chosen for the U.S. digital terrestrial HDTV system, may also be used.

The MPEG-2 system-level functions specify how the sources for video, sound, and data are combined into a *transport stream* (TS) for applications in lossy and erred channels such as storage media and the technical broadcast media. The MPEG-2 system level also contains processing that comprises a *program stream* for applications in relatively error-free transmission channels (e.g., ATM networks) and interactive multimedia systems. The overall system approach can be regarded as a combination of multiplexing at two different layers. In the first layer, multiplexing transport packets from one or more *packetized elementary streams* (PESs) sources form single-program transport bit streams. In the second layer, many single-program TSs are combined to form a system of programs. The *program-service information* (PSI) streams contain the information relating to the identification of programs and the components of each program. This data is transmitted in the form of tables, such as the *program association table* (PAT) and the *program map table* (PMT), to

Table 4.3
MPEG-2 Codes Situating Table

Profiles, Levels	Simple Profile (no-B-Frames, 4:2:0)	Main Profile (4:2:0)	SNR Scalable Profile (4:2:0)	High-Profile (Other Tools; 4:2:0)
High-1920 level (HDTV)	—	MP@HL	—	HP@HL
High-1440 level (HDTV)	—	MP@H14L (Europe)	SSP@H14L	HP@H14L
Main-level (SDTV)	SP@ML	MP@MP	—	HP@ML
Low-level (CIF)	—	MP@LL	—	—

Notes: Combinations marked with (—) are not defined; SP: simple profile, MP: main profile, SSP: SNR scalable profile, HP: high profile, LL: low level, ML: main level, HL: high-1920 level, H14L: high-1440 level.

automatically configure the decoder and allow it to reconstitute the transmitted information in signal components (video, audio, or data) for each compressed program. Figure 4.5 shows the basic structure of the TS multiplex in MPEG-2. The output of the TS is organized as a TDM baseband digital signal.

The TS packets have a length of 188 bytes (Figure 4.6) so that the payload is 184 bytes. The fixed length of transport packets is used to ease introduction of error control methods in the transmission process. The *packet identifier* (PID) identifies what kind of service is in each 188-byte TS packet, for example, PAT, PMT, or compressed video.

4.4   The Digital TV Satellite Transmission Problem

A typical digital TV satellite transmission link is shown in Figure 4.4. The task of the satellite modulator/demodulator is to transmit reliably the MPEG-2 TS over the satellite channel, comprising the transponder, the uplink equipment, the uplink and downlink paths, and the downlink receiving equipment.

4.4.1   Low Error Rate for MPEG-2 with a Low C/N

MPEG-2 demultiplexing and decompression processes are highly sensitive to bit errors. As a result, an extremely low BER is required to provide acceptable

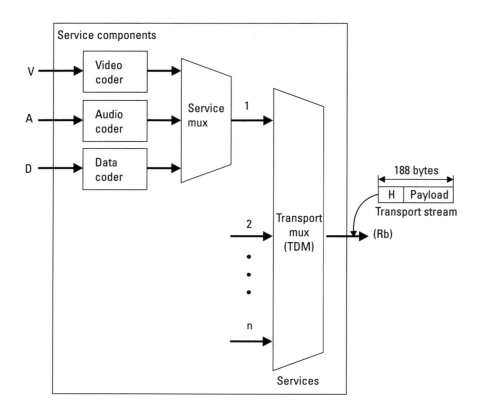

Figure 4.5 MPEG-2 source coding and multiplexing into a TS.

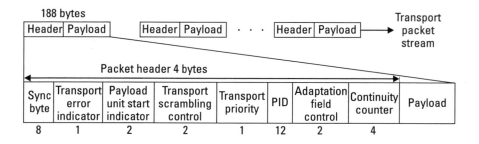

Figure 4.6 MPEG-2 TS syntax.

service to the end user. The maximum BER that can be tolerated is 10^{-10} to 10^{-11} and corresponds to a *quasi-error-free* (QEF) quality target. On the receiver side, the satellite transmission system must be capable of working with a relatively low C/N at the receiver. For this reason, powerful error-correction coding (concatenated coding) associated with the modulation scheme must be used.

4.4.2 Variety of Signal and Transponder Bandwidths

From one application to another the bandwidth of the signal transmitted over the satellite channel may vary widely as a function of the satellite transponder bandwidth and of the transponder occupancy. In practice, transponder bandwidths range from 24 to 72 MHz. In addition, depending on the application, one may wish to trade off receiver antenna size against the information bit rate and therefore vary the degree of error-correction coding. The satellite transmission system should be flexible with regard to different transponder bandwidths and must allow different receiving antenna sizes.

4.4.3 Modulation and Coding Scheme

Given a specific satellite, an important tradeoff for the design of a satellite TV transmission system is between the selection of a receiver antenna size (which determines gain and figure of merit G/T) and the selection of a modulation/coding scheme to achieve a cost-effective balance between the following factors:

- The transmission bit rate;
- Receiver ground terminal cost and performance;
- The space segment needed for the service (transponder bandwidth per program and output RF power).

4.5 DVB-S Standard (ETS 300 421 Transmission System)

4.5.1 Overview

The European Standard ETS 300 421 [11] describes the modulation and channel coding system for satellite digital multiprogram SDTV and HDTV services to be used for primary and secondary distribution in the FSS and

BSS Ku bands (Figure 4.7). In other words, this modulation and coding scheme is intended for DTH services for consumer IRD, as well as for collective antenna system [*satellite master antenna TV* (SMATV)] and cable television head-end stations, with a likelihood of remodulation.

The input to the transmission scheme of the DVB-S standard comprises packets with a size of 188 bytes each according to the MPEG-2 TS specifications (and not accidentally equal to 4×47-byte ATM packet payloads). A unified MPEG-2 digital time-multiplexed TS may contain eight or more SDTV services, with associated audio, auxiliary audio services, CA data, and auxiliary data services such as teletext or Internet connectivity. A single VHS-quality movie can be transmitted at an information bit rate of 1.5 Mbps, a news or general entertainment TV program at 3.4–4 Mbps, and live sports at 4–6 Mbps. The encoding rate required for any MPEG-2 broadcast varies according to the decisions made by each program service provider.

After the desired content stream is formatted, the following processes are applied to the 188-byte MPEG-2 TS (Figure 4.8):

1. Randomization for energy dispersal;

2. Reed-Solomon outer coding;

3. Interleaving;

4. Convolutional inner coding;

5. Baseband filter shaping;

6. Modulation.

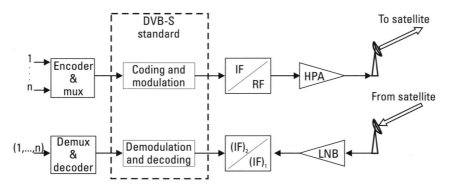

Figure 4.7 System block diagram and scope of the DVB-S standard.

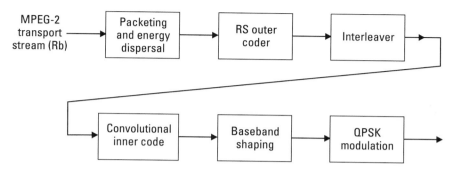

Figure 4.8   Baseband and modulator block diagram used in the DVB-S standard.

4.5.2   Randomization for Energy Dispersal

To achieve a regular spectrum shape of transmitted signals to facilitate clock-recovery in the receiver, the data at the output of the MPEG-2 multiplexer is bit-by-bit randomized. The scrambling *pseudorandom binary sequence* (PRBS) is synchronized to a frame of eight MPEG-2 transport packets (Figure 4.9) delimited by two-inverted MPEG-2 sync bytes (SYNC1). During the transmission of the sync bytes the scrambler is disabled.

4.5.3   Error-Correction Coding and Decoding

Error-correction encoding (and decoding) is now applied on the randomized transport packets. To obtain the required performance with a limited decoding complexity, a concatenated code comprising an inner convolutional code and an outer Reed-Solomon code is used [12]. With typical C/N values of

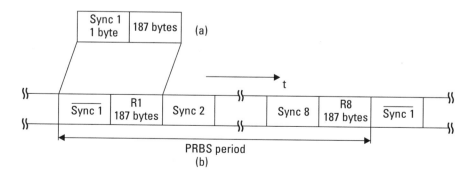

Figure 4.9   Randomization process: (a) MPEG TS packet and (b) randomized sequence (R1/R8).

4–8 dB at the demodulator input, convolutional coding using a Viterbi decoder at the receiver offers the best performance on the AWGN channel, which is a good model for the satellite channel.

To allow for a variable C/N range (and thus a choice in the receiver antenna size), variable-rate convolutional codes can be used. These variable rates are made possible by using punctured codes. From the convolutional mother code of rate 1/2, it is possible to obtain rates of 2/3, 3/4, 5/6, and 7/8. A convolutional code of rate 3/4 is shown in Figure 4.10 using a puncture matrix.

The other code rates can be obtained using the matrix puncture pattern shown in Table 4.4.

In the concatenated coding scheme adopted in the standard, the residual errors at the output of the Viterbi decoder are grouped in bursts of 50–100 bits. The outer code must be able to combat effectively with these burst errors. Using an interleaver that spreads the error burst in the code word over a number of random-erred code words does this.

In the DVB-S specification, a Reed-Solomon code with code words of 204 bytes (188 information bytes and 16 redundant bytes) is used; this is capable of correcting up to eight random byte errors. This means that each MPEG-2 transport packet is encoded into one R-S code word. The different R-S decoding algorithms are capable of efficiently reducing an input BER of 10^{-3}–10^{-4} into an output BER of 10^{-10}–10^{-11}. The overall encoding and decoding scheme used is shown in Figure 4.11.

Let r_{RS} be the Reed-Solomon code rate and r_C be the convolutional code rate. It is possible to write the following expression

$$R_C = \frac{R_b}{r_{RS} \cdot r_C}, \text{ Mbps} \tag{4.4}$$

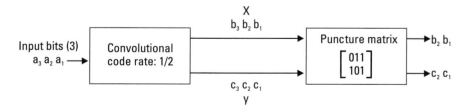

Figure 4.10 Convolutional code rate of 3/4.

Table 4.4
Variable Code Rates Using Puncture Codes

Code Rate	Matrix Puncture Pattern
1/2	$x: \begin{bmatrix} 1 \\ 1 \end{bmatrix}$ $y:$
2/3	$x: \begin{bmatrix} 10 \\ 01 \end{bmatrix}$ $y:$
3/4	$x: \begin{bmatrix} 011 \\ 101 \end{bmatrix}$ $y:$
5/6	$x: \begin{bmatrix} 01011 \\ 10101 \end{bmatrix}$ $y:$
7/8	$x: \begin{bmatrix} 0101111 \\ 1010001 \end{bmatrix}$ $y:$

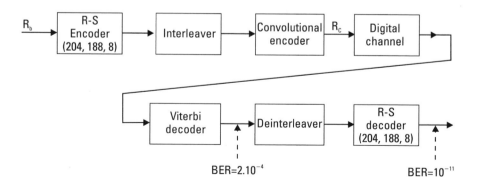

Figure 4.11 Encoding-decoding process in the DVB-S standard.

where R_C is the coded bit transmission rate and R_b is the MPEG-2 TS transmission bit rate, both in megabits per second. It should be noted that (4.4) shows an implicit increment in transmission bandwidth.

The coding gain G_C (the difference in the C/N for an objective target BER between an uncoded system and a coded system) that is achieved in this concatenated encoding-decoding scheme depends on the convolutional code rate and can be calculated using the required value of E_b/N_0 without coding, for 10^{-11} output BER using QPSK (13.5 dB), and the corresponding values of E_b/N_0 reported in [13]. G_C values are shown in Table 4.5. Therefore, for example, to get 10^{-11} output BER with DVB-S rate 1/2 coding (as in the first entry in Table 4.5), one needs input $E_b/N_0 = (13.5 - 9.0) = 4.5$ dB. The coded bits are Gray-mapped in the QPSK constellation as shown in Figure 4.12.

4.5.4 Baseband Shaping

The coded bits are filtered at baseband to generate a square-root raised cosine spectrum defined by

$$H(f) = \begin{cases} 1 & ; |f| \leq \dfrac{R_S}{2}(1-\alpha) \\[4mm] \left\{ \dfrac{1}{2} + \dfrac{1}{2}\sin\left[\dfrac{\pi}{2R_S} \cdot \left(\dfrac{R_S - |f|}{\alpha} \right) \right] \right\}^{1/2} & ; \dfrac{R_S}{2}(1-\alpha) \leq |f| \leq \dfrac{R_S}{2}(1+\alpha) \\[4mm] 0 & ; |f| > \dfrac{R_S}{2}(1+\alpha) \end{cases}$$

(4.5)

Table 4.5
Concatenated Coding Gain G_C Versus Convolutional Code Rate in the DVB-S Standard

r_c	1/2	2/3	3/4	5/6	7/8
G_C(dB)	9.00	8.50	8.00	7.50	7.10

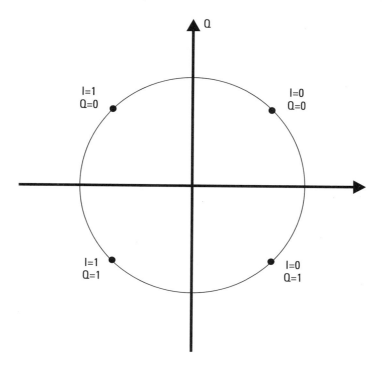

Figure 4.12 QPSK constellation.

where R_S is the symbol rate and α is the roll-off factor ($0 \leq \alpha \leq 1$). A typical value is $\alpha = 0.35$. This frequency response is used in the transmitter and receiver sides to reduce intersymbol interference. The theoretical spectrum is shown in Figure 4.13.

The bandwidth for the equivalent linear passband-shaping filter for infinite attenuation is

$$B = (1 + \alpha)R_S, \text{MHz} \tag{4.6}$$

The relation between the baud rate R_S and the coded bit transmission rate R_C in a multilevel transmission system with M statistically independent and equiprobable symbols is

$$R_S = \frac{R_C}{\log_2 M}, \text{Msymb/second} \tag{4.7}$$

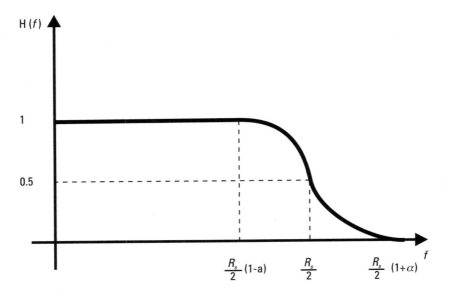

Figure 4.13 Baseband shaping filter frequency response.

Substituting (4.7) into (4.6) obtains

$$B = \frac{R_C}{\log_2 M} \cdot (1 + \alpha), \text{ MHz} \qquad (4.8)$$

4.5.5 Modulation

To achieve very high-power efficiency without excessively penalizing the spectrum efficiency, the DVB-S standard specifies the use of coherent QPSK modulation.

The spectral efficiency η_S can be defined as

$$\eta_S = \frac{R_C}{B_T}, \text{ bps/Hz} \qquad (4.9)$$

Substituting (4.8) in (4.9) obtains

$$\eta_S = \frac{\log_2 M}{(1 + \alpha)}, \text{ bps/Hz} \qquad (4.10)$$

and for QPSK with $M = 4$

$$\eta_S = \frac{2}{(1 + \alpha)} \quad (4.11)$$

Let B_T be the transponder bandwidth for the transmission of a group of digital compressed signals. Using (4.9), it is possible to write

$$\eta_S = \frac{R_C}{B_T}, \text{ bps/Hz} \quad (4.12)$$

Using (4.4), (4.10), and (4.12), one obtains

$$R_b = \frac{\log_2 M}{(1 + \alpha)} \cdot r_{RS} \cdot r_C \cdot B_T, \text{ Mbps} \quad (4.13)$$

if B_T is expressed in megahertz, as it is usually.

Equation (4.13) represents the satellite downlink capacity to transmit digital compressed signals using the DVB-S standard. This presumes that there is sufficient signal strength, an aspect we will cover after Examples 4.2 and 4.3.

Example 4.2

A 36-MHz transponder is going to be used in digital TV transmissions according to standard DVB-S. The FEC code rate is 3/4 and the roll-off factor is 0.2. Determine the number of programs that can be time-multiplexed in the MPEG-2 TS if the MPEG-2 encoder has an output bit rate of 4 Mbps (video plus audio). Consider an additional 1.8 Mbps per program for data, SI, CA, synchronism, and header.

Solution. Using (4.13), obtain

$$R_b = \left(\frac{2}{1 + 0.2} \right) \cdot \left(\frac{188}{204} \right) \cdot \left(\frac{3}{4} \right) \cdot 36 = 41.47 \text{ Mbps}$$

If N represents the number of programs in the time-multiplexed digital signal, then

$$R_b = N(4 + 1.8) = 6 \cdot N$$

$$N = \frac{41.47}{5.8} = 7.15$$

which can be approximated to seven programs.

Example 4.3

In the Example 4.2, determine the coded bit rate (maximum bit rate) in the satellite channel.

Solution.   Using (4.12) and (4.11), obtain

$$R_C = \frac{2}{1.2} \cdot 36 = 60 \text{ Mbps}$$

which represents a relatively high bit rate.

The BER expressed as a bit error probability, P_b, for coherent QPSK, is

$$P_b = Q\left(\sqrt{\frac{2E_b}{N_0}} \right) \qquad (4.14)$$

where E_b/N_0 (energy useful bit relative to the noise power per hertz) is the key performance parameter and $Q(.)$ is the special function [14]

$$Q(x) = \frac{1}{\sqrt{2\pi}} \int_x^{\infty} e^{-t^2/2} dt \qquad (4.15)$$

which can be approximated by

$$Q(x) = \frac{1}{x\sqrt{2\pi}} \cdot e^{-x^2/2} \qquad (4.16)$$

when $x > 3$.

Example 4.4

A QPSK signal is transmitted by satellite. Raised cosine filtering is used for which the roll-off factor is 0.35 and a BER of 10^{-11} is required. For the satellite downlink the free-space loss is 205.4 dB, the receiving G/T is 12 dB/K, the additional loss equals 3.5 dB, and the rain margin is 4 dB. The transponder bandwidth is 24 MHz. The DVB-S standard is used with a convolutional code rate of 3/4 and 8 dB of coding gain. Calculate the following:

1. The maximum information bit rate for the downlink satellite channel;
2. The required EIRP.

Assume 0.5 dB for the uplink noise contribution.

Solution.

1. Using (4.13),

$$R_b = \frac{2}{1+0.35} \cdot \frac{188}{204} \cdot \frac{3}{4} \cdot 24 = 24.57 \text{ Mbps}$$

2. For $P_b = 10^{-11}$ and QPSK modulation, the E_b/N_0 is 13.5 dB, which can be verified using (4.16). The information bit rate R_b, in decibels per bits per second, is

$$R_b = 10 \log\left(24.57 \cdot 10^6\right) = 73.9 \text{ dB.bps}$$

Using now (3.60), the required EIRP for the transmitting satellite is

$$\text{EIRP} = 4 - 12 + 205.4 + 3.5 + 73.9 + 0.5 - 8 + 13.5 - 228.6 = 52.2 \text{ dBW}$$

When program providers purchase time on a satellite, in effect they are primarily paying for bandwidth. Therefore, if a programmer wanted to transmit three TV programs via a transponder, it should use less bandwidth than a service that transmitted six. However, the bandwidth of a transponder is finite, and therefore an upper limit is placed on the symbol rate (typically between 28 and 30 Msymb/second). By reducing the amount of redundant information sent along with the actual data (higher values of the convolutional code rate), the number of programs can be increased. However, this then means that satellite channel errors are harder to correct and that the downlink ground stations must be able to receive a higher modulated carrier power level (i.e., use a larger dish) to receive the expected quality programming via the transponder. Hence, making the right choice of capacity versus coding is critical.

4.6   DVB Service Information (DVB-SI)

DVB has arranged an open service information system to come with the DVB-S signal. It is to be used by the IRD to tune to a particular service selected by the system user. The MPEG-2 PSI information allows the IRD to automatically configure itself, and the DVB-SI information enables the IRD to tune to a particular service(s), as grouped into categories with relevant program information. The DVB-SI standard implements the service delivery model from components and services. Components are associated to a compressed signal of video, audio, or data. Service is associated to a program that contains, for example, one video component, two audio components, and one data component. Several services are jointly multiplexed and sent via specific transponders on a satellite. A user can then select a subset of services (bouquet) that can be transmitted by one or several transponders. DVB-SI also provides the elements needed to produce an *electronic program guide* (EPG). Relevant schedule information and descriptions of the programs are broadcast over the same link with the signal and thus provide the EPG directly to users. DVB-SI is based on four tables, plus other optional tables. They contain descriptors that outline the characteristics of the service or event. The four tables are described as follows:

- *Network information table (NIT):* The NIT groups together services belonging to a network provider. It contains tuning information to be used during IRD setup and signals a change in tuning information. (The system adapts automatically to the error characteristics of the channel. The broadcaster can use different channel encoder code rates that the receiver will automatically lock on the transmitted parameters by a rapid trial-and-error process.)

- *Service descriptor table (SDT):* The SDT lists the names and parameters associated with each service in a MPEG multiplex.

- *Event information table (EIT):* The EIT contains information about the current transport and optionally covers others TSs that the IRD can receive.

- *Time and data table (TDT):* The TDT is used to update the IRD's clock and calendar.

Additional optional SI tables are described as follows:

- *Bouquet association table (BAT):* The BAT provides a means of grouping services that the IRD might use to present available services to the end user. A particular service can belong to more than one bouquet.

- *Running status table (RST):* The RST provides detailed information about the current program. Unlike other SI tables, the RST is transmitted only once, while others are transmitted repeatedly.

- *Stuffing table (ST):* The ST is used to replace, invalidate, or modify other DVB-SI tables.

4.7 Conditional Access

Satellite TV systems use CA methods to avoid unauthorized access to the offered programming. The insertion of CA into IRDs is an essential factor in the economics of broadcast services.

CA systems are based on two basic techniques: scrambling and encryption. Scrambling is required to render the transmitted signal meaningless to the IRD unequipped with a means of descrambling the received transmission. The ability of an IRD to descramble is conveyed to the receiver in the form of a key or covert digital number, and an encryption process is required to make this key secret.

In DVB, there are only a few packet types that must be transmitted without scrambling. Obviously, these include part of the *service information* (SI) stream such as the PAT and the NIT. These data streams need to be transmitted without scrambling so that any DVB-compliant receiver can access this specific information.

The system to support DVB-CA requires a database, known as the *subscriber management system* (SMS), to manage the subscribers, their addresses, and program requirements. The program requirements are sent to the IRD using an appropriately structured message known as the *entitlement management message* (EMM). The CA system timing and synchronization, together with the current encryption key, are sent in the *entitlement control message* (ECM). These terms apply to the system adopted by the DVB standard.

DVB uses two basic CA procedures: Simulcrypt and Multicrypt. An IRD implemented with Simulcrypt would only work on a network that is set up for this CA arrangement. In contrast, an IRD implemented with Multicrypt is able to work with a common interface (DVB-CI) to allow an open-system approach to the normally proprietary CA system architecture. There are a number of different companies providing DVB-compatible CA systems (e.g., Nagravision, Irdeto, and Seca), so when DBS providers start a service, they have many options from which to choose. It is important to carefully consider the CA system when the transmission standard is not compatible with DVB-S. (Consider, for example, VideoGuard, which is used with DirecTV.)

The key used for the encryption process is usually transmitted over the satellite link along with the programming and other information. This is the most efficient and successful means of delivery of the electronic key. Users, identified by a smart card that also has a unique key, ask and pay for wanted programming via a terrestrial return channel using a voice-grade modem. Most smart card serial interfaces operate in the 9,600–38,400-bps range. The key used to scramble the program changes over time. The serial communication between CI and the smart card occurs with a burst of data every few seconds.

Scrambling of the appropriate bit streams is performed at the uplink site. The MPEG-2 packets are encrypted by the usual techniques, based on a common key known to both the scrambling and decryption devices. When a scrambled packet arrives, it is passed through the CI, which takes the key obtained from the smart card and uses it to turn the packet payload back into an MPEG-2 transport payload that can be processed by the rest of the system.

4.8   North America's Standards

In the United States, Mexico, and Canada, DVB-S has not been fully adopted, and other standards are used. They are described here.

Digicipher is a GI standard for satellite program distribution. The first version, Digicipher I, is still used today by main providers. Digicipher I was really the first digital compression scheme that was commercially available and so a large number of program providers still use it.

Digicipher II (DCII) is GI's most recent standard; it uses MPEG-2 encoding for video. However, almost all of the remaining technology used in this standard is exclusive to DCII. For example, DVB-S uses Musicam for audio, whereas DCII uses Dolby AC 3. Despite the same video standard, DVB-S and DCII signals are totally incompatible, and no receiver can currently receive both.

Although both DCII and DVB-S have the same performance, the differences, especially in the area of the transmission standard, prevent DCII receivers from receiving DVB-S signals and vice versa. The main attributes of DCII and DVB-S are described as follows:

- Both standards use QPSK as the modulation scheme.

- DCII uses a mother convolutional code rate of 1/3 instead of the 1/2 rate used in DVB-S.

- DCII uses a Butterworth pulse-shaping filter instead of the square-root-raised cosine pulse-shaping filter used in DVB-S.

- DCII uses standard MPEG-2/MP@ML encoding for video (the same as DVB-S) and Dolby AC 3 for encoding audio (Musicam for DVB-S).

Although there are a lot of DCII video-compatible products in North America, DVB-S is also very popular there because it is cheaper to produce DVB-S encoders and receivers than DCII encoders and receivers. By the beginning of 2000, more than 1,000 DVB-S–compatible services were able to be received from U.S. companies. EchoStar, ExpressVu, and Microspace use the same encoder and receiving equipment to transmit digital video. All share the same symbol rate of 20 Msymb/s and transmit using DVB-S. DirecTV does not use the standard DVB-S because it went on the air before the aforementioned standard was ratified. This standard is called DSS, and, much like DCII, it uses MPEG-2 encoding for video but has a different audio encoding system and different service information packets. Additionally, the DSS service uses variable convolutional code rate values (1/2, 2/3, 6/7) depending on the transponder bandwidth. The Reed-Solomon outer code used is (146, 130, 8). The DSS standard is incompatible with DVB-S receivers and vice versa.

4.9 Transponder Access

Each transmitting Earth station accesses a nonregenerative satellite transponder in a single channel (analog TV program) per carrier operation as shown in Figure 4.14. This allows each satellite transponder to operate in saturation with its output power at its maximum value. As shown in Figure 4.14, each transmitting Earth station uplinks one analog TV program.

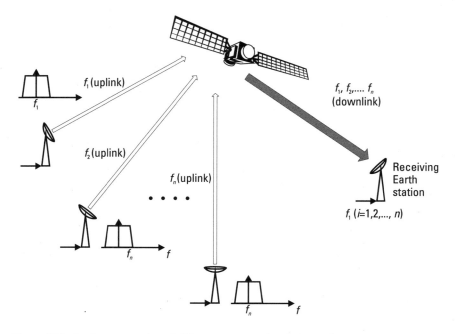

Figure 4.14 Analog access to satellite nonregenerative transponder.

Various transmitting Earth station signals could be joined together in one uplink feeder ground station, but there is only one uplink per satellite transponder.

The access method used for digital satellite TV can be the same as that used for the analog case where the transponder is a nonregenerative one, and a single modulated carrier is used in the whole transponder bandwidth (see Figure 4.15).

The main advantages of this approach are described as follows:

1. This access method is compatible with the nonregenerative technology payload of most current geostationary satellites in operation and under construction.

2. The single-carrier mode allows the operation in saturation for the TWTA onboard the satellite transponder achieving maximum power output, without intermodulation.

3. The RF/L-band-IF section of the analog satellite TV ground reception system can be used for digital ground terminals. Some

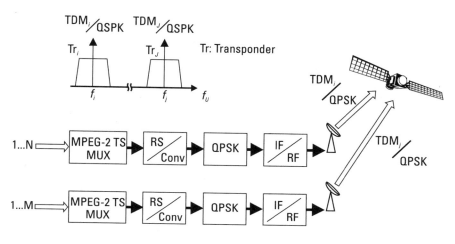

Figure 4.15 Transponder access in a digital satellite TV system using a TDM baseband signal from the output of the MPEG-2 transport multiplex and modulating a QPSK single carrier.

hardware changes occur if the digital band is different from the analog band, but these are minor and do not require a great deal of development to overcome.

4. Common packages of program services can use the same CA and forward correction systems, thereby economizing on the overall bandwidth and transmission bit rate requirements.

The major drawback of this kind of single-carrier/transponder access is that all program streams must be put together in a transmitting Earth station as illustrated in Figure 4.15. A single program provider cannot transmit directly to the satellite transponder; instead, it must be included in the TDM baseband signal to allow the MCPC mode associated with the operation in saturation. The European Skyplex system [15] provides a means to overcome this problem by allowing different TV broadcasting centers to independently access the satellite payload and to send their program contents to home receivers. Figure 2.9 outlined the Skyplex system architecture. Each broadcasting center uses a small and low-cost ground station transmitting at a low bit rate corresponding to its own program content in a FDMA basis. Aboard the satellite, the different low-bit rate signals are received, regenerated, and multiplexed to form a single high-bit rate digital stream using TDM.

The technical approach adopted by Skyplex is related to the partitioning of the DVB-S standard to minimize the amount of onboard processing and the broadcaster uplink station size. The main features of the Skyplex architecture are described as follows:

- The generation of the transport packets is performed at a ground facility at each broadcasting station and operated independently for the associated uplink TS. These single-channel TS multiplexers are similar to the conventional MPEG-2 TSs but (usually) transmit one digital program at a low bit rate. The single-channel multiplexers contain most of the complexity of the MPEG-2 TS multiplexer.

- The scrambling function and the outer Reed-Solomon encoding is performed on-ground for each single-channel MPEG-2 multiplexer, independently for each uplink TS. The Reed-Solomon encoder output is QPSK-modulated, fed to the RF transmitting section of the ground station, and uplinked to the corresponding satellite.

- The interleaving, convolutional inner encoding, and QPSK modulation are performed onboard after multiplexing, creating a high-bit rate stream using TDM. The QPSK-modulated carrier is then downlinked to the ground in the proper RF frequency band and with the required power level.

To have a negligible effect on the overall transmission quality, it is necessary to ensure a BER of 10^{-5} (or lower) at the onboard QPSK demodulator output. It is important to note that there is no error correction on the uplink.

The uplink design is very similar to that of the downlink. In analogy with (3.57), it is possible to write the following uplink design equation:

$$M_R = (\text{EIRP})_T + (\text{G / T})_S - L_b$$
$$-\Sigma L - B_T - (\text{C / N})_{o,U} + 228.6, \text{dB} \tag{4.17}$$

where M_R is the fade margin, $(\text{EIRP})_T$ is the EIRP for the transmitting Earth station, $(\text{G/T})_S$ is the satellite G/T, L_b is the free-space loss (calculated using the uplink frequency band), B_T is the transponder bandwidth, and $(\text{C/N})_{o,U}$ is the required C/N for the uplink. The term ΔN_U (uplink noise contribution) is neglected for obvious reasons. It is assumed that the transponder is operating at saturation.

The radio propagation conditions in the atmosphere affect the uplink and downlink in a different way. Rain increases attenuation and antenna noise temperature for the downlink as it is expressed in (3.58). In the uplink case, it should be assumed that rain only increases attenuation in the transmission path. This is because the satellite antenna is pointed toward a "hot" Earth, and this noise temperature (about 300K) tends to mask any additional noise induced by rain attenuation. Then, the rain margin for the uplink is

$$M_R \geq A_R, \text{dB} \tag{4.18}$$

Example 4.5

An uplink facility is located at 49° N, 2.33° E; operates at 14 GHz; and transmits TV signals to a satellite located at 13° E. The flux density required to saturate the transponder (nonregenerative) is –80 dW/m². Determine the minimum antenna diameter for the transmitting ground station if the TWTA has a power output of 1 kW. Coupling losses between the transmitter output and the antenna's input amount to 2.5 dB. Assume 60% for antenna efficiency and clear-sky conditions in the transmission path.

Solution. The distance between the ground station and the satellite can be calculated using (3.11). Then

$$d = 4.264 \cdot 10^4 \cdot \sqrt{1 - 0.296\cos(13° - 2.33°) \cdot \cos(49°)} = 38,356.2 \text{ km}$$

Now, using (3.28),

$$(\text{EIRP})_T = -80 + 10\log\left[4\pi\left(38,356.2 \cdot 10^3\right)^2 \right] = 82.67 \text{ dBW}$$

If the TWTA power output is 30 dBW, then the antenna gain is

$$G_T = 82.67 - 30 + 2.5 = 55.17 \text{ dBi}$$

and the corresponding diameter is

$$D = \frac{0.0214}{\pi} \cdot \sqrt{\frac{10^{5.517}}{0.6}} = 5.04 \text{ m}$$

Example 4.6

Repeat Example 4.5, now considering a rain attenuation of 5.5 dB for 99.9% service availability in the average year.

Solution. Using (4.18), the margin can be selected as 6 dB. The required antenna gain, under rain conditions and using the same TWTA power output (1 kW), is now

$$G_T = 55.17 + 6 = 61.17 \, \text{dBi}$$

and the corresponding antenna diameter (60% efficiency) is

$$D = \frac{0.0214}{\pi} \cdot \sqrt{\frac{10^{6.117}}{0.6}} = 10.07 \text{m}$$

If the same antenna gain is kept as in Example 4.5, then the new TWTA power output must be

$$P_{TWTA} = 82.67 - 55.17 + 6 = 33.5 \, \text{dBW}$$
$$= 2.24 \text{ kW}$$

It is important to underline in the last solution that some form of uplink power control is necessary to compensate for rain fades in the uplink. Thus, the Earth-station HPA must have enough reserve power to meet the fade margin requirement.

For the digital case, (4.17) can be transformed to

$$M_R = (\text{EIRP})_T + (\text{G}/\text{T})_S - L_b - \Sigma L$$
$$-R_b - \left(E_b / N_0\right)_{o,U} + 228.6, \text{dB}$$
(4.19)

The coding gain G_C is not included in (4.19), because the error control is performed at the receiver ground terminal for both nonregenerative and Skyplex-like regenerative transponders.

Example 4.7

Using the Skyplex architecture, determine a proper combination between the TWTA power output and the antenna diameter for the transmitting Earth station. Use the following parameters:

- Uplink frequency: 14 GHz;

- Free-space loss: 207.4 dB;

- Satellite (G/T): 1.30 dB/K;

- Bit rate (corresponding to a single program for each Earth station): 6 Mbps;

- Rain attenuation: 5.5 dB (99% service availability in the average year);

- Additional losses: 2.1 dB;

- BER (at the output of the onboard QPSK demodulator): 10^{-5}.

Solution.   Using (4.18), the margin can be chosen as 6 dB. For QPSK and a BER of 10^{-5}, the (E_b/N_0) value is 9.6 dB. The bit rate in decibels per bits per second is

$$R_b = 10 \log\left(6 \cdot 10^6\right) = 67.78 \text{ dB.bps}$$

Substituting the numerical values into (4.19), obtain

$$(\text{EIRP})_T = 6 - 1.3 + 207.4 + 2.1 + 67.78 + 9.6 - 228.6 = 62.98 \text{ dBW}$$

Assuming a 50-W TWTA (17 dBW), the antenna gain is

$$G_T = 62.98 - 17 = 45.98 \approx 46 \text{ dBi}$$

and the corresponding diameter is

$$D = \frac{0.0214}{\pi} \cdot \sqrt{\frac{10^{4.6}}{0.6}} = 1.75 \approx 1.80 \text{m}$$

Comparing Examples 4.5 and 4.7, the main advantage of Skyplex is the elimination of the large central uplink facility, resulting in a remarkable operating cost reduction for individual broadcasters.

4.10   Digital SMATV Systems

Satellite television broadcasting, although primarily focused on DTH reception, requires signal distribution in receiving community installations commonly called SMATVs. SMATV networks are domestic TV systems using low-cost consumer technology and simple design methods without regular performance control like those used by CATV networks. SMATV networks are usually used for serving single buildings or condominiums. SMATV systems combine satellite and terrestrial TV channels efficiently so that all residents in a multiple-unit dwelling can access the available programming by means of a single cable network. SMATV networks also use fiber optics when coaxial cable is difficult and expensive to implement.

The adoption of one unique modulation scheme optimized for both satellite and cable is practically impossible. In fact, the satellite channel is basically power-limited and nonlinear, and it does not suffer from stringent bandwidth limitations, whereas cable channels are linear and allow relatively high C/N ratios but are sharply band-limited and affected by echoes and other distortions.

The modulation and coding standard for CATV networks has been defined in the DVB standard known as DVB-C. To allow maximum transparency and receiver commonality, the same baseband processing (i.e., randomization, Reed-Solomon coding, and interleaving) of the satellite system has been adopted. However, high-level quadrature modulation methods (i.e., 16-QAM, 32-QAM, or 64-QAM, with 15% roll-off) have been introduced to allow transport of typical satellite bit rates on 6- or 8-MHz cable channels. To increase spectrum efficiency, the convolutional inner code is not adopted.

Two distribution techniques have been investigated in Europe to assess the suitability of the DVB standards and systems for use in SMATV installations. These studies have led to the definition of the ETSI standard ETS 300 473 (DVB-CS).

There are two approaches:

- Method A: Transparent distribution of the DVB-S QPSK signals by simple frequency conversion in the extended superband (230–470 MHz) available in current installations in Europe (SMATV-S) and/or at the satellite first IF (SMATV-IF) [16]. This method allows roughly eight channels to be distributed.

- Method B: Error correction and remodulation from DVB-S (QPSK) to DVB-C format (16-QAM, 32-QAM, or 64-QAM) and distribution in 6- or 8-MHz channels.

Distribution method B—although more complex and expensive than method A because of the need of remodulation from QPSK to QAM (16, 32, or 64) of each satellite channel—allows better spectrum efficiency. In the extended superband (230–470 MHz) currently used in SMATV installations in Europe, up to 34 satellite channels can be distributed, with 8-MHz channel spacing, corresponding to about 170 SDTV programs.

4.11 Further Developments in Transmission Techniques

Over the past few years various studies have been conducted for future digital broadcasting systems to support HDTV and multimedia services using the 12- and 20-GHz frequency bands [17, 18]. Japanese proponents generally lead these studies, and their system is known as *integrated services digital broadcasting* (ISDB).

The following modulation and channel-coding schemes have been selected for these systems:

- Trellis-coded modulation 8PSK (TCM-8PSK) and Reed-Solomon outer code;

- QPSK and concatenated Reed-Solomon and punctured convolutional coding;

- BPSK, convolutional code, (code rate 1/2) and Reed-Solomon code.

TCM-8PSK is used as the main modulation scheme (code rate 2/3) to permit the high transmission capacity for broadcasting one or two HDTV programs on a single satellite transponder. The Reed-Solomon code used is the same as the DVB-S standard and is compatible with the output packets of the MPEG-2 TS multiplexer (188-byte length). The (E_b/N_0) requirement is about 6.1 dB for a BER of 10^{-11} (QEF) and $0 \le R_b/B_T \le 1.5$ [19]. The required-information bit rates for HDTV and SDTV are 22 Mbps and 6 Mbps, respectively. These results are derived from picture-quality tests conducted by the Association of Radio Industries and Business (ARIB) in Japan [20].

In addition to TCM-8PSK and QPSK, BPSK (in association with the corresponding channel coding) can also be used to provide higher service availability. Broadcasters can choose the most suitable modulation and coding scheme for their services from among these various schemes according to the C/N expected in the targeted receive coverage zone. Figure 4.16 from [20] shows the tradeoff achieved.

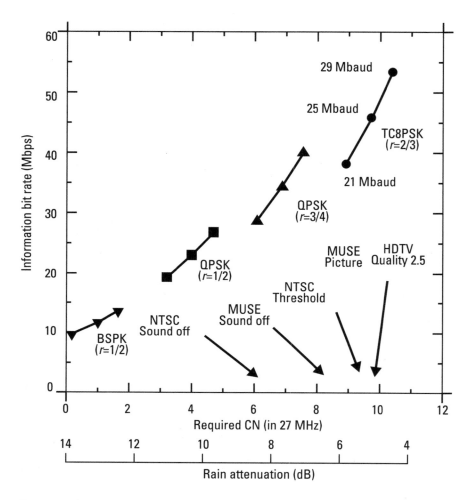

Figure 4.16 ISDB modulation type information bit rate versus required C/N (from [20]).

The ISDB scheme is unique in its time multiplexing of the above-mentioned modulation methods into the same stream. For instance, part of the stream on a given transponder can be in TCM 8PSK, and some of the remainder could be BPSK, depending on traffic needs.

Example 4.8

Determine the downlink margin increase when a broadcaster switches from TCM-8PSK to QPSK with a convolutional code rate of 1/2. Both schemes use the same Reed-Solomon code as the DVB-S standard. Assume a roll-off factor of 0.25. The transponder bandwidth is 36 MHz. The other downlink parameters remain the same.

Solution.   The information bit rate for TCM-8PSK is

$$R_b = \frac{\log_2 8}{1 + 0.25} \cdot \frac{188}{204} \cdot \frac{2}{3} \cdot 36 = 53.08 \text{ Mbps}$$

and in decibels per bits per second is

$$R_b = 10 \log(53.08 \cdot 10^6) = 77.25 \text{ dB.bps}$$

The information bit rate for QPSK and a convolutional code rate of 1/2 is

$$R_b = \frac{2}{1 + 0.25} \cdot \frac{188}{204} \cdot \frac{1}{2} \cdot 36 = 26.54 \text{ Mbps}$$

and in decibels per bits per second is

$$R_b = 10 \log(26.54 \cdot 10^6) = 74.24 \text{ dB.bps}$$

The E_b/N_0 required for TCM-8PSK is 6.1 dB and for QPSK is 4.5 dB (see Table 4.5), both for a BER of 10^{-11}.

Then, using (3.60), the increase in downlink margin (favorable to QPSK) is

$$\Delta M_R = (6.1 + 77.25) - (4.5 + 74.24) = 4.61 \text{ dB}$$

Subsequently, the QPSK scheme used in Example 4.8 has a higher protection to rain fades, and service availability will be higher than in the TCM-8PSK scheme.

Another important improvement to protect satellite transmissions from rain fading is the use of turbo codes [21] associated, in this case, with the low-bit rate modulation and coding scheme (QPSK and BPSK). Turbo codes are a new class of convolutional code using a parallel concatenation of two recursive systematic convolutional codes; their performance, in terms of BER, is close to the Shannon limit ($E_b/N_0 = -1.6$ dB).

Example 4.9

Repeat Example 4.8, now using a turbo code (code rate 1/2 and E_b/N_0 = 1.24 dB for a BER of 10^{-11}) instead of the inner convolutional code associated with the QPSK scheme.

Solution. The information bit rate with the QPSK scheme, now using a turbo code, is the same as before—that is, 26.54 Mbps (74.24 dB.bps). The increase in downlink margin will now be

$$\Delta M_R = (6.1 + 77.25) - (1.24 + 74.24) = 7.87 \text{ dB}$$

Another key issue is the service continuity to allow what is called *graceful degradation* under deteriorating atmospheric conditions. Analog TV signal degrades progressively as a function of the degradation of the reception conditions. Digital TV tends to degrade rapidly once the threshold of the C/N operating point is reached, and it is possible to go from virtual error-free reception to complete loss of picture decoder operation (the so-called brick-wall effect).

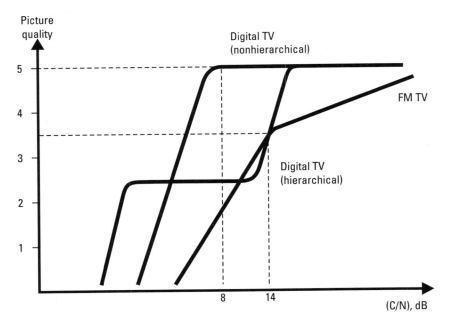

Figure 4.17 Hierarchical channel coding performance.

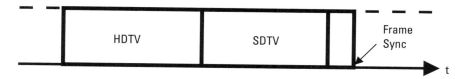

Figure 4.18 Hierarchical transmission frame.

Hierarchical transmission has been investigated to solve this problem without increasing total transmission power. During severe rain attenuation, the receiver automatically switches from the HDTV components to the SDTV components, which are transmitted simultaneously in a time-multiplexed fashion. The switching criterion can be related to the received power or to the bit error ratio of the HDTV component. For a given satellite transmitted power, this approach allows an extension of the service continuity and hence a reduction in the service outage time. Figure 4.17 shows the degradation of picture quality in the cases of a conventional analog system, a nonhierarchical digital system, and a hierarchical digital system.

Figure 4.18 shows a typical hierarchical transmission frame. The simultaneous transmission and reception method shown in Figure 4.19 is one

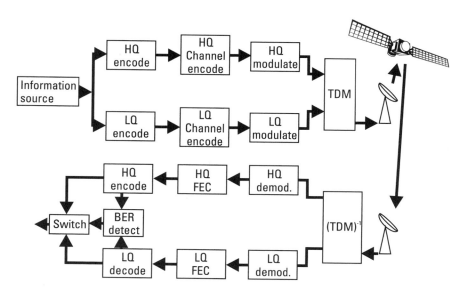

Figure 4.19 Hierarchical transmission system.

alternative combining a *high-quality layer* (HQL) for HDTV and a *low-quality layer* (LQL) for SDTV.

References

[1]   Drury, G. M., "Digital Video Broadcasting by Satellite," in *Satellite Communication Systems*, B. G. Evans (ed.), London: IEE, 1999, pp. 397–455.

[2]   Taub, H., and D. Schilling, *Principles of Communications Systems*, New York: McGraw-Hill, 1986.

[3]   Pritchard, W., and M. Ogata, "Satellite Direct Broadcast," *Proc. IEEE*, Vol. 78, No. 7, July 1990, pp. 1116–1140.

[4]   ATSC Standard, "Modulation and Coding Requirements for Digital TV (DTV) Applications over Satellite," Doc. A/80, July 17, 1999.

[5]   Shikora, T., "MPEG Digital Video-Coding Standards," *IEEE Signal Processing Magazine*, Vol. 14, No. 5, September 1997, pp. 82–100.

[6]   Elbert, B. R., *The Satellite Communication Applications Handbook*, Norwood, MA: Artech House, 1997, pp. 169–204.

[7]   ISO/IEC 13818-1: "Coding of Moving Pictures and Associated Audio: Systems," November 1993.

[8]   ISO/IEC 13818-1: "Coding of Moving Pictures and Associated Audio: Video," November 1993.

[9]   ISO/IEC 13818-3: "Coding of Moving Pictures and Associated Audio: Audio," November 1993.

[10]   Noll, P., "MPEG Digital Audio Coding," *IEEE Signal Processing Magazine*, September 1997, pp. 59–81.

[11]   ETSI, "Digital Broadcasting Systems for Television, Sound and Data Services: Framing Structure, Channel Coding and Modulation for 11/12 GHz Satellite Services," Draft PR ETS 300 421, June 1994.

[12]   Reed, I., and X. Chen, *Error-Control Coding for Data Networks*, Norwell, MA: Kluwer Academic Publishers, 1999.

[13]   Comminetti, M., et al, "The European System for Digital Multi-Program Television by Satellite," *IEEE Trans. on Broadcasting*, Vol. 41, No. 2, June 1995, pp. 49–62.

[14]   Proakis, J., *Digital Communications*, New York: McGraw Hill, 1995.

[15]   Elia, C., and E. Colzi, "Skyplex: Distributed Up-Link for Digital Television Via Satellite," *IEEE Trans. on Broadcasting*, Vol. 42, No. 4, December 1996, pp. 305–313.

[16]   Stephenson, D. J., *Guide to Satellite TV*, Boston, MA: Newnes, 1997, pp. 290–313.

[17] Kawai, N., S. Namba, and S. Yamaki, "Performance of Multimedia Broadcasting Through ISDB Transmission System," *IEEE Trans. on Broadcasting*, Vol. 42, No. 3, September 1996, pp. 151–158.

[18] Combarel, L., et al., "HD-SAT Modems for the Satellite Broadcasting in the 20-GHz Frequency Band," *IEEE Trans. on Consumer Electronics*, Vol. 41, No. 4, November 1995, pp. 991–999.

[19] Katoh, H., et al., "A Flexible Transmission Technique for the Satellite ISDB System," *IEEE Trans. on Broadcasting*, Vol. 42, No. 3, September 1996, pp. 159–166.

[20] Minematsu, F., et al., "Transmission System for Multimedia Services in Satellite Broadcasting Channels," *IEEE Trans. on Consumer Electronics*, Vol. 44, No. 3, August 1998, pp. 556–563.

[21] Berrou, C., and A. Glavieux, "Near-Optimum Error Correcting Coding and Decoding: Turbo-Codes," *IEEE Trans. on Communications*, Vol. 44, No. 10, 1996, pp. 1261–1271.

5

Satellite TV Reception System: Architecture, Technology, and Performance

5.1 Receiver System Architecture

The receiver system architecture is composed of the following elements (see Figure 5.1):

- The receiving antenna;
- The LNB;
- The IRD;
- The coaxial cable that interconnects the LNB and the IRD.

The receiving antenna is typically a parabolic dish and comprises a feed, a reflector, and a mechanical support called the mount. The receiving antenna is pointed to a satellite that transmits a number of modulated carriers from specific transponders.

The output of the receiving antenna's feed passes those modulated carriers that have the same polarization of the LNB. The LNB amplifies and downconverts in frequency the aforementioned carriers in order for them to be adequately transmitted by conventional low-cost coaxial cables. For example, in C-band systems the modulated carriers are downconverted from

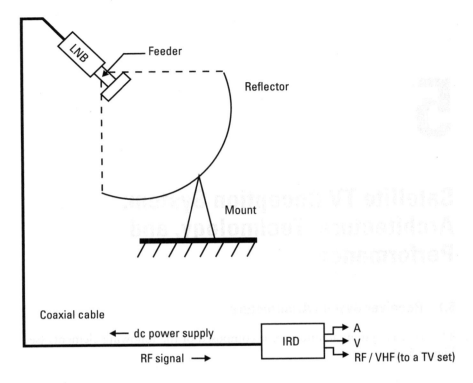

Figure 5.1 Receiver system basic architecture.

3.7–4.2 GHz to 950–1,450 MHz. This frequency range, from the LNB's output, is called the "first IF" of the receive system. It should be remembered that attenuation in coaxial cable rises as frequency increases and that consumer-grade cable should not be used above 3 GHz. The dc power supply to the LNB is fed from an IRD using the same coax but in an opposite direction to the RF signals coming from the receiving antenna.

The IRD is a typical superheterodyne receiver that translates one specific modulated carrier (in the first IF range) to a fixed second IF. The encrypted baseband signal is obtained from the detector output (analog or digital) and decoded by the decoder unit. The clear baseband signal is usually remodulated in the VHF-UHF band, using an analog TV standard (e.g., NTSC, PAL, or SECAM), and it is fed to the antenna input of a conventional TV set.

5.2 Antennas

The typical gain needed in the antenna of a satellite receiving system is more than 20 dBi, and, for that reason, the parabolic antenna is the most popular antenna used. For a C-band system the prime focus parabola (symmetric parabola) is typically used, and for Ku-band systems the offset parabola is the most widely used antenna. Sections 5.2.1–5.2.5 summarize the main antenna parameters.

5.2.1 Gain

The gain of an antenna is the ratio of the power radiated (or received) per unit solid angle by the antenna in a given direction to the power radiated (or received) per unit solid angle by an isotropic antenna fed with the same power. The gain is a maximum in the direction of maximum radiation (boresight) and, for a parabolic antenna, it is expressed as

$$G(\text{dBi}) = 10 \cdot \log\left[\eta\left(\frac{\pi D}{\lambda}\right)^2\right] \qquad (5.1)$$

where:

  G: Antenna gain, in decibels relative to isotropic antenna;

  η: Antenna efficiency (typical values: from 55% to 70%);

  D: Circular aperture diameter, in meters;

  λ: Wavelength in free-space conditions, in meters.

5.2.2 Half-Power Beamwidth

The half-power beamwidth is the angle between the directions in which the gain falls to half its maximum value. This angle is called BW° and a practical formula for parabolic antennas is (Figure 5.2)

$$\text{BW}° = 70 \cdot \frac{\lambda}{D} \qquad (5.2)$$

In a direction θ near the boresight, $(0 \le \theta \le \text{BW}°/2)$, the value of the gain is given by

$$G(\theta) = G - 12 \cdot (\theta/\text{BW}°)^2, \text{ dBi} \qquad (5.3)$$

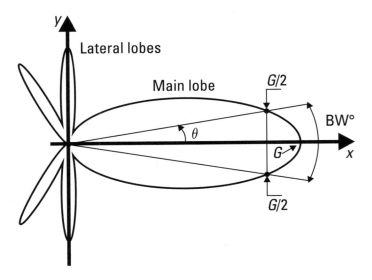

Figure 5.2 Antenna beamwidth.

Combining (5.1) and (5.2), obtain

$$G(\text{dBi}) = 10 \cdot \log\left[\eta \cdot \left(\frac{70\pi}{\text{BW}°}\right)^2\right] \qquad (5.4)$$

5.2.3 Polar and Cross-Polar Antenna Radiation Pattern

The polar antenna radiation pattern indicates the variations of gain with directions for a given polarization. The cross-polar radiation pattern is the antenna radiation pattern that corresponds to the orthogonal polarization and should be zero for an ideal antenna operating in the opposite orthogonal polarization. For a real receiving antenna with one kind of polarization, there is not an absolute isolation to the opposite orthogonal polarization, and it will be a residual undesired interference. The *cross-polar isolation* (XPI) is a typical measure of the orthogonal polarization rejection for a receiving antenna.

5.2.4 Antenna Noise Temperature

The antenna noise temperature is an indirect external noise power measure in the detector input and in the *IF amplifier* (IFA) bandwidth of the superheterodyne receiver. Then, if there is impedance matching among superheterodyne sections, it is possible to write (Figure 5.3)

$$N_{\text{ext}} = kT_A B, \text{W} \tag{5.5}$$

where N_{ext} represents the external noise power, k is the Boltzmann's constant, and B is the IF amplifier's equivalent noise bandwidth.

The antenna noise temperature depends on the antenna radiation pattern and external noise sources as sky noise (cosmic plus atmospheric noise) and ground noise and can be calculated by the following expression:

$$T_A = \frac{1}{4\pi} \cdot \int_0^{2\pi} \int_0^\pi T_b(\theta, \varphi) g(\theta, \varphi) \sin d\,\theta\varphi, \text{K} \tag{5.6}$$

where $T_b(\theta, \varphi)$ is the external noise source noise temperature (in Kelvin) and $g(\theta, \varphi)$ represents the gain variations with direction. The coordinates (θ, φ) represent spherical coordinates where the antenna is located at the origin.

The expression (5.6) is a theoretical one and is cumbersome for engineering purposes. Equation (5.7) can reproduce approximately the values of antenna noise temperature in the C band (4 GHz) reported in [1].

$$T_A = \frac{77}{D(\text{m})} + \frac{454}{EL^\circ(\text{deg})}, \text{K} \tag{5.7}$$

A general guideline for antenna noise temperature in the Ku band (12-GHz band) is

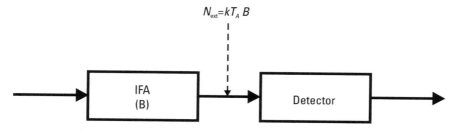

Figure 5.3 Antenna noise temperature.

$$T_A \leq \begin{cases} 35K \text{ ;EL}°\geq 30° \\ 50K \text{ ;EL}° < 30° \end{cases} \qquad (5.8)$$

for clear-sky conditions (no rain).

5.2.5 Antenna Pointing Techniques

A variety of aiming technologies have been developed to allow receiving antennas to be pointed at one or more geostationary communication satellites.

Antennas serving for the reception of geostationary satellite transmission are divided in two main groups depending on their mount: biaxial antennas (azimuth-elevation mount type) and uniaxial antennas (polar mount type).

In the case of biaxial antennas, both the azimuth (AZ°) and elevation (EL°) must be adjusted to the direction of maximum radiation (Figure 5.4).

Satellite signals transmitted with linear polarization are oriented relative to the equatorial plane. The feed on an AZ-EL-mounted parabola is oriented with the local horizontal plane; therefore, the reception of linearly

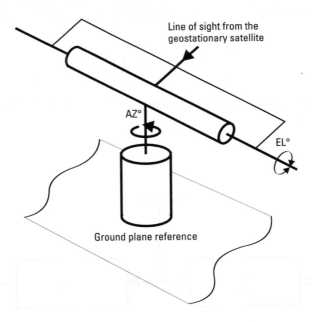

Figure 5.4 AZ-EL mounting.

polarized signals from satellites to the east or west of the Earth station site requires some polarization adjustment of the feed for optimal reception (Figure 5.5).

The polarization angle value ψ can be calculated [2]:

$$\psi = -\arctan\left(\frac{\sin \Delta\varphi}{\tan \theta}\right) \tag{5.9}$$

If the polarization angle is negative (looking from the satellite), the correction is clockwise, and it is counterclockwise if it is positive. AZ-EL–mounted parabolic antennas may receive satellite signals transmitted with circular polarization without concern for the feeder position correction.

The polar mount is used in antenna systems when needed to sweep a segment of the geostationary orbit. Its advantage is that it can do this with only one rotational movement, thus using only one actuator. Because the geostationary arc is curved when viewed from anywhere besides the equator —even though the actuator allows only one degree of rotational freedom— this kind of mount has inherent errors when it is tracking the geostationary

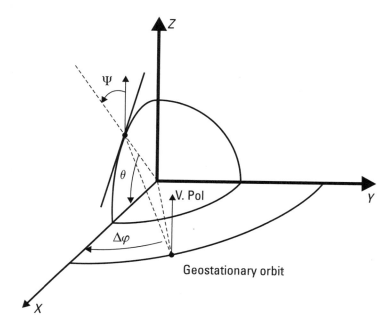

Figure 5.5 Polarization angle for vertical polarization.

orbit. By analyzing Figure 5.6, corresponding to the polar mount, it is possible to write

$$I° = \theta° \ (\text{Earth station latitude}) \qquad (5.10)$$

$$D° = \arctan\left(\frac{\sin\theta}{6.61 - \cos\theta}\right) \qquad (5.11)$$

where I° is the inclination angle and D° is the declination angle.

The modified polar mount is used to minimize the aforementioned tracking errors in a range of

$$|\Delta\varphi| \le \arccos(0.1513\cos\theta),\ |\theta| < 813° \qquad (5.12)$$

in the geostationary arc, and it introduces a tilt B° (also known as Birkill's factor) for the polar axis given by [3]

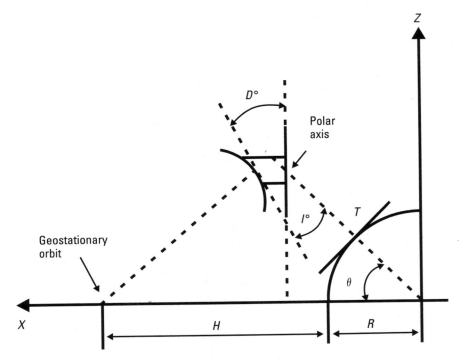

Figure 5.6 Polar mount.

$$B° = \arctan\left(\frac{6.61 \cdot \sqrt{1 - 0.023\cos^2\theta}}{\sin\theta}\right)$$

$$- \arctan\left(\frac{6.61 \cdot 1 - 0.15\cos\theta}{\sin\theta}\right) \tag{5.13}$$

Then, the new inclination angle value is given by

$$I° = \theta° + B° \tag{5.14}$$

and the declination angle is now

$$D° = 90° - \arctan\left(\frac{6.61 \cdot \sqrt{1 - 0.023\cos^2\theta}}{\sin\theta}\right) \tag{5.15}$$

The modified polar mount is illustrated in Figure 5.7.

Example 5.1

A receiving Earth station is located in 22° N, 80° W, and should track the geostationary orbit from 125° W to 101° W. Determine the modified polar mount angles.

Solution. Using (5.12), obtain

$$|\Delta\varphi| = 24° < 81.9°$$

The polar axis tilt is

$$B° = \arctan\left(\frac{6.61 \cdot \sqrt{1 - 0.023\cos^2(22°)}}{\sin(22°)}\right)$$

$$- \arctan\left(\frac{6.61 \cdot 1 - 0.15\cos(22°)}{\sin(22°)}\right) = 0.49°$$

and

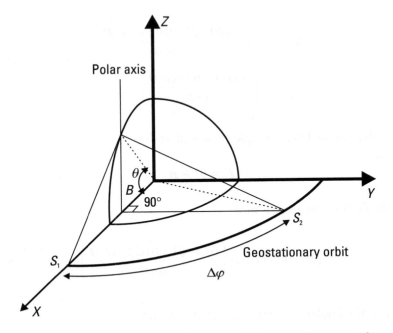

Figure 5.7 Maximum polar mount sweeping range in the geostationary arc: from S_1 to S_2 (hypothetical satellite).

$$I° = 22° + 0.49° \cong 22.5°$$

The declination angle is then

$$D° = 90° - \arctan\left(\frac{6.61 \cdot \sqrt{1 - 0.023 \cos^2(22°)}}{\sin(22°)}\right) = 3.276°$$

5.3 Low Noise Block

The basic block diagram of an LNB, also known as an outdoor unit, is shown in Figure 5.8.

The LNB provides two main functions: amplifying the RF signals by means of an LNA and downconverting the input RF frequency range to the first IF range. The RF input is coupled to the LNA by means of a waveguide transition and an isolator to reject signals coming from the high-stability *local oscillator* (LO), or *dielectric resonator oscillator* (DRO) into the RF input. A typical value of the first IF range is 950–1,450 MHz or 950–2,150 MHz, very common in Europe for ASTRA satellites

In Ku-band systems the local oscillator frequency value (f_{LO}) is lower than the frequency value of the input RF signals (f_{RF}) coming from the geostationary satellite (low-side injection). Then

$$f_{1/IF} = f_{RF} - f_{LO}, \text{ MHz} \qquad (5.16)$$

and the output signal spectrum is not inverted relative to the input signal.

In C-band systems the situation is opposite (high-side injection); that is,

$$f_{1/IF} = f_{LO} - f_{RF}, \text{ MHz} \qquad (5.17)$$

and the output signal spectrum is now inverted relative to the input signal.

Example 5.2

The input RF range in C-band systems is 3.7–4.2 GHz, and the DRO frequency used in the LNB is 5.15 GHz. Determine the following:

1. The LNB spectrum output in the first IF;
2. The image channel spectrum.

Solution. Remembering some basics in a superheterodyne receiver, it is possible to write

$$f_{1/IF} = f_{im} - f_{LO}, \text{ MHz}$$

where f_{im} is the image channel. The results are displayed in Figure 5.9.

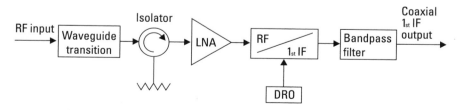

Figure 5.8 LNB and downconverter.

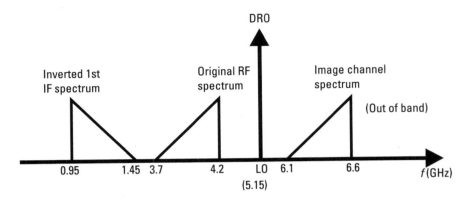

Figure 5.9 Solution to Example 5.2.

Example 5.3

The input RF signal spectrum for ITU-2 Ku-band broadcast satellite service ranges from 12.2 to 12.7 GHz in the downlink. The LNB used in a satellite TV reception system has a DRO that is tuned at 11.25 GHz. Determine the following:

   1. The LNB spectrum output in the first IF;
   2. The image channel spectrum.

Solution.   Now, it is possible to write

$$f_{1/IF} = f_{LO} - f_{im}, \text{ MHz}$$

and the results are displayed in Figure 5.10.

The noise characteristics of an LNB can be given as noise temperature (C-band systems) or noise figure (Ku-band systems), and the LNA section determines it. Typical values are listed as follows:

   • C-band LNB: T_{LNB} = 25–40K;
   • Ku-band LNB: F_{LNB} = 0.7–1.3 dB.

The above figures actually represent the highest noise level that occurs anywhere within the passband of the LNB.

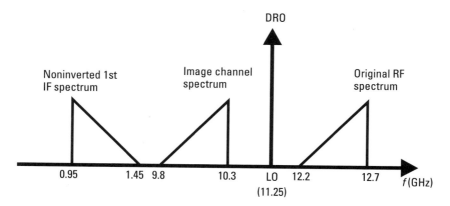

Figure 5.10 Solution to Example 5.3.

The conversion gain (G_{LNB}) is defined as the power gain measured between the output (first IF) and the input (RF). The consumer-grade LNB commonly produces 50–65 dB, fairly constant over the passband of the LNB. The gain specification for any LNB is important for ensuring that the signal arriving at the second IF input of the IRD is within the range of values in dBm (sensitivity) recommended by the IRD manufacturer, taking into consideration the maximum allowed length of coaxial used between the LNB output and the IRD input.

Also available is a product called a universal Ku-band LNB that can switch electronically between the 10.7–11.7-GHz and 11.7–12.75-GHz frequency subbands to provide complete coverage of the entire Ku-band frequency range. Via the coaxial, the IRD sends up a switched 13/17V dc to switch between orthogonal polarizations or a 22-kHz tone to select one of two available local oscillators (9.75 GHz or 10.6 GHz) inside the LNB.

The most critical performance parameter for an oscillator used in LNB is the long-term frequency drift. The frequency stability of an oscillator is affected by aging in the components that determine the frequency of oscillation. Also, temperature changes affect the frequency-controlling components and cause the oscillator to drift. These frequency changes are characterized in terms of the temperature coefficient of the components and how these affect the overall temperature stability of the oscillator. The temperature coefficient is the fractional change in the parameter per degree change in temperature. The fractional change is usually given in percent or in parts per million (ppm), and the temperature change is given in degrees Celsius (°C). DROs used in LNBs have better than 10 ppm/°C (10 kHz/GHz per °C change in temperature) temperature coefficients. An LNB used in analog systems may thus have an LO stability of ±500 kHz. Digital systems usually employ a phase-locked LNB with ±15 kHz of frequency stability.

Phase noise is another important parameter in the LNB specification. Phase noise is the result of short-term instability of the DRO. Excessive LNB phase noise can cause the IRD to reproduce a picture that contains erroneous components. An ideal oscillator with an infinite quality factor (Q) would exhibit single-line spectra at its fundamental frequency. Practical oscillators, however, have a finite Q due to losses in the components used to realize the oscillator's feedback mechanism. This results in phase-noise accumulation at the offset from the carrier. The undesired energy is distributed symmetrically around the carrier, exponentially decaying as the frequency departs from the carrier on both sides. The shape of the distributions is based on the character-istics of the active device and the Q of the resonant circuit. One way to char-acterize the performance of the phase noise of an oscillator is to quantify the residual noise energy at a specified offset from the carrier. A common parameter used to define this residual energy is the one-sided *power-spectral density* (PSD), which is the noise power contained in a 1-Hz bandwidth at a particular frequency offset from the carrier. Typical values are –60 dBc/Hz at 1 kHz, –95 dBc/Hz at 10 kHz, and –115 dBc/Hz at 100 kHz. The unit dBc/Hz means decibels relative to the carrier, in a 1-Hz bandwidth.

The LNB is a high-technology component. The uses of Gallium Arse-nide (GaAs) FETs (GaAsFETs) and *high electron mobility transistors* (HEMTs) have been key factors in achieving a good expected performance of the LNB and its low-cost implementation. Most digital systems use a com-pact combination of LNB and feed (LNBF).

5.4 Integrated Receiver and Decoder

The IRD, also known as the indoor unit or set-top box, is a superheterodyne receiver, as shown in Figure 5.11 for the analog case.

The *baseband processor* (BBP) block permits the audiovisual signal decoding according to the encryption algorithm used by the satellite service provider.

Some of the most important parameters in identifying IRD perform-ance are listed as follows:

- Input operating frequency spectrum, compliant with the LNB output;

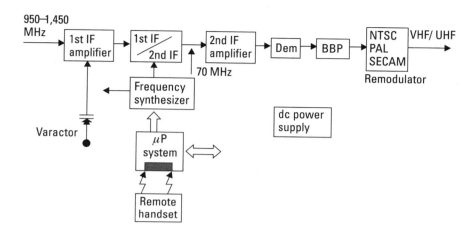

Figure 5.11 Analog IRD block diagram.

- Sensitivity (typical values range from –30 to –60 dBm);
- Input impedance (75Ω, unbalanced);
- Noise figure (a typical value is 12 dB);
- Second IF bandwidth (typical values range from 20–30 MHz);
- FM detector threshold.

Figure 5.12 shows the block diagram of a digital reception system compliant with the DVB-S standard [4]. The receiver mainly consists of a tuner, including an IF *surface acoustic wave* (SAW) filter; an analog QPSK demodulator; and the digital channel decoder that is a QPSK processor device. In this system, the first IF signal from the LNB (950–1,450 MHz or 950–2,150 MHz) is mixed down (high-side injection) to a second IF with a typical center frequency of 480 MHz. In current designs the image frequency is 960 MHz above the desired channel. Two IF amplifiers are used between the SAW filter to compensate the relative high-filter losses (20 dB). Since DVB-S uses several information bit rates, the bandwidth of the tuner is variable (e.g., 24 and 36 MHz). After the second IF filtering, the signal is I/Q-demodulated to a baseband signal between dc and 12 MHz (or 18 MHz) and a level of, for example, 1 Vpp.

The I and Q outputs of the QPSK demodulator are each digitized for further processing in the QPSK processor. Each signal must be sampled at the same instant to preserve the relative phase relationship between them

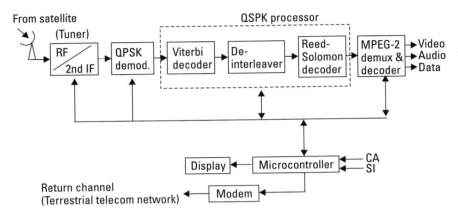

Figure 5.12 Digital satellite receiver block diagram (CA: conditional access; SI: information).

with dual 5- or 6-bit *analog-to-digital converters* (ADCs) at a minimum rate equal to twice the signal's maximum symbol rate. In the DirecTV system, for example, a sampling rate of 40 MHz is used. Other systems operate with sampling rates of 60 MHz and can reach as high as 90 MHz.

The digital baseband circuitry of the QPSK processor processes incoming samples from the ADCs. These sampled digital signals first pass through a pair of matched filters before synchronization, timing recovery, and error-correction operations are performed (Viterbi decoding, convolutional deinterleaving, and FEC with the Reed-Solomon algorithm). The QPSK processor output digital bit stream then passes to an MPEG-2 demultiplexer and decoder block. The demultiplexer recovers a single MPEG-2 TV program from the MPEG-2 TS. Finally, the MPEG-2 program stream is sent to a MPEG-2 decoder to get the audiovisual baseband signals and auxiliary data and remodulated as Channel 3 (VHF) for reception by a TV set tuned to that frequency.

Industry pressure is mounting to increase consumer acceptance by lowering the cost of digital IRDs [5–7]. To meet these industry demands, a direct-conversion tuner—to simplify the front-end circuitry by directly translating the first IF signals (950–1,450 MHz or 950–2,150 MHz) to baseband I/Q output signals—has been developed. The new circuit completely eliminates the need for a 480-MHz second IF and its associated circuitry. Instead of a SAW filter (setting the system's selectivity), the filtering on the direct-conversion system is performed in baseband frequency with two separate filters on I- and Q-paths having the same transfer function. Differences between these filters lead to I/Q impairments; therefore, their amplitude response and group delay should be the same. The phase imbalance should be less than 3°, and mismatch gain between I and Q channels over the entire frequency tuning range should be equal to or less than 0.5 dB. These filters avoid aliasing on ADC and keep as much adjacent channel power as possible away from the second amplifier stage. The other stages are mainly the same as for the double conversion system. This kind of tuner accepts carrier levels of –70 to –20 dBm and has a dynamic range of more than 50 dB. This large *automatic gain control* (AGC) range is necessary to accommodate rainfall attenuation effects, differences in coaxial cable lengths, and less-than-perfect DBS parabolic-dish-antenna alignment.

Most digital IRDs are programmed by the manufacturer to receive specific DVB-S transmission parameters belonging to specific compressed digital TV programs from just one satellite (bouquet). This initial setup includes the satellite transponder's center frequency and polarization format, as well as the service provider's symbol and convolutional code rates. For example, DVB-S transmission parameters over a typical 24-MHz transponder on EchoStar I are listed as follows:

- Symbol rate: 20 Msymb/second;
- FEC rate = 3/4;
- CA = Nagravision.

Once installed, the digital IRD will tune automatically to the manufacturer-programmed "default transponder" and access the EPG, SI, and CA data that it needs before it can begin delivering audiovisual signals to the TV set. For example, 12.177 V SR 23.00 FEC 2/3 means that the transmission is centered at 12.177 GHz, uses vertical polarization for the downlink, uses a symbol rate of 23 Msymb/second, and has FEC 2/3.

5.5 Coaxial Cable

Typical coaxial cables used in satellite TV reception systems (analog and digital) have 75-ohm impedance. Some attenuation values are shown in Table 5.1.

The specific attenuation α (dB/m) at frequency f can be interpolated using the approximate expression

Table 5.1
Attenuation Versus Frequency Values for Typical Coaxial Used in Satellite TV Reception Systems

Designation	Transversal Section Geometry	Specific Attenuation (dB/m)	
		100 MHz	1,450 MHz
RG-59	⊙	0.1	0.36
RG-6	⊙	0.09	0.28
RG-11	⊙	0.08	0.23

$$\alpha = \alpha_0 \cdot \left(\frac{f}{f_0} \right)^n , \, dB/m \qquad (5.18)$$

where α_0 is the specific attenuation at frequency f_0 and n is a real-valued exponent.

5.6 Receiver Performance

5.6.1 Overall Satellite TV Reception System Noise Temperature

The overall thermal noise model of a satellite TV reception system is shown in Figure 5.13. The input thermal noise in the receiver is considered at the output of the dish's feed, and it will be assumed that the feed loss can be neglected. However, it should be remembered that feed could introduce more thermal noise into the reception system. For example, if feed loss has a value of 0.1 dB at the operating frequency, it means an additional noise of $290(10^{0.01} - 1) = 6.75K$ at the feed's input. If this fact cannot be neglected, the feed loss must be added to the "other losses" term in the margin equation, and the additional noise temperature must be added to the antenna noise temperature T_A for clear-sky conditions. Following the above example, the new antenna noise temperature will be $10^{-0.01} \cdot 36.75 = 35.9K$ at the feeder's output, instead of 30K without considering the feed loss (see Appendix D).

As the thermal noise model does not depend on the carrier frequency value but on the carrier bandwidth (kTB), the downconversion process that

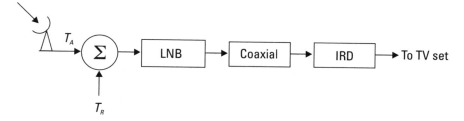

Figure 5.13 Thermal noise model of a satellite TV reception system where T_A represents the antenna noise temperature (external noise) and T_R is the receiver noise temperature (internal noise).

occurs in the LNB can be ignored for noise temperature calculations. It is now possible to consider the following parameters:

- T_{LNB}: LNB noise temperature (in Kelvin);
- G_{LNB}: LNB conversion gain (in decibels);
- T_{COAX}: Coaxial noise temperature (in Kelvin);
- L_{COAX}: Coaxial loss (in decibels);
- T_{IRD}: IRD noise temperature (in Kelvin).

The above parameters are evaluated in the proper carrier frequency. Using the well-known formula for calculating the overall receiver noise temperature, it is possible to write

$$T_R = T_{LNB} + \frac{T_{COAX}}{10^{G_{LNB}/10}} + \frac{T_{IRD}}{10^{G_{LNB}/10} \cdot 10^{-L_{COAX}/10}} \tag{5.19}$$

and, with some algebraic manipulations,

$$T_R = T_{LNB} + \frac{\left(T_{COAX} + T_{IRD} \cdot 10^{-L_{COAX}/10}\right)}{10^{G_{LNB}/10}}, K \tag{5.20}$$

Example 5.4

Determine the receiver noise temperature for a C-band system if $T_{LNB} = 25K$ and $G_{LNB} = 60$ dB. Use a 5-m coaxial RG-11 cable between the LNB. The IRD has a noise figure of 12 dB.

Solution. The coax loss must be calculated for the worst case (1,450 MHz). Then, using Table 5.1,

$$L_{COAX} = 0.23 \cdot 5 = 1.15\,dB$$

The value of T_{COAX} can be calculated using the conversion formula between noise temperature (T) and noise figure (F). Remember that $F_{COAX} = L_{COAX}$; then,

$$T_{COAX} = 290 \cdot \left(10^{0.115} - 1\right) = 87.92\,K$$

and

$$T_{IRD} = 290 \cdot \left(10^{1.2} - 1\right) = 4{,}306.19\,K$$

Substituting the above values in (5.20),

$$T_R = 25 + \frac{87.92 + 10^{0.115} \cdot 4{,}306.19}{10^6} = 25.01\,K$$

From the above example it is possible to conclude that, for engineering purposes,

$$T_R = T_{LNB}, K \tag{5.21}$$

and, in a equivalent way,

$$F_R = F_{LNB}, dB \tag{5.22}$$

5.6.2 Interference Analysis

In almost all past practical cases, satellite system performance is based on noise limitations. Nonetheless, there are a growing number of special situations where interference is a major problem, such as in the case of C-band systems in North America, where satellites are spaced 2° or 3° from each other in the geostationary arc and may use the same frequency plan and polarization for several transponders. The analysis that follows is an engineering approach to the numerical estimation of interference. At its conclusion, a few practical comments useful for downlink design are added.

Interference can be regarded as an unwanted RF power introduced at the receiver system. If the interferer power I is small and can be considered as an equivalent noncoherent Gaussian noise source, it is possible to extend the overall C/N in the following way:

$$(C/N)^{-1} = (C/N)_U^{-1} + (C/N)_D^{-1} + (C/I)^{-1} \qquad (5.23)$$

where C/I is the carrier-to-interference ratio. Equation (5.23) can also be written as

$$(C/N)^{-1} = (C/N)_D^{-1} \Delta N_U + (C/I)^{-1} \qquad (5.24)$$

Using some algebraic manipulations,

$$(C/N)_D = (C/N) \cdot \Delta N_U \cdot \Delta I \qquad (5.25)$$

where ΔI is defined by

$$\Delta I = \frac{1}{\left[1 - \dfrac{C/N}{C/I} \right]}; C/I \geq C/N \qquad (5.26)$$

and can be considered as an additional degradation factor in the system's performance.

Example 5.5

Determine the numerical value of ΔI (in decibels), if C/N = 14 dB and:

1. C/I = 40 dB;
2. C/I = 15 dB.

Solution. Using (5.26), obtain

$$\Delta I = 10 \cdot \log \left[1 - 10^{(14-40)/10} \right] = 0.011 \, \text{dB}$$

$$\Delta I = 10 \cdot \log\left[1 - 10^{(14-15)/10}\right] = 6.87 \, \text{dB}$$

Substituting (5.25) in the downlink margin (3.57)

$$M_R = \text{EIRP} + G / T - L_b - \Sigma L - B$$
$$- (C / N)_0 - \Delta N_U - \Delta I + 228.6, (\text{dB}) \tag{5.27}$$

and, for the digital case,

$$M_R = \text{EIRP} + G / T - L_b - \Sigma L - R_b$$
$$- (E_b / N_0)_0 - \Delta N_U - \Delta I + G_C + 228.6, (\text{dB}) \tag{5.28}$$

There are four common types of interference in satellite TV reception systems:

1. Cross-polarization interference (X_{POL});
2. *Adjacent channel interference* (ACI);
3. *Terrestrial interference* (TI);
4. *Orbital interference* (OI).

X_{POL} usually comes from the same satellite transmitting the intended channel on the frequency allocations corresponding to the adjacent transponders (the lower and upper ones in frequency) with orthogonal polarization relative to that it is tuned for wanted reception. (This presumes that a standard staggered channel frequency plan is used. If the frequency plan is unstaggered, there will be only one channel on the opposite polarization.) X_{POL} transponders have a residual signal at the output of the antenna feed because of the nonideal cross-polar antenna radiation patterns of both the ground receive and spacecraft transmit antennas. A typical example is shown in Figure 5.14.

Also, the cross-polar antenna radiation pattern effect (even transponders) is shown.

X_{POL} is controlled in the planning stage of satellite TV networks by adequately selecting radio channel frequency spacing, bandwidth, and polarization. For most broadcast situations the user's receive dish, with as-installed alignment, is 10–15 dB worse than the X_{POL} of the carefully aligned spacecraft transmit antenna. Thus, an indirect measure for X_{POL} rejection is the

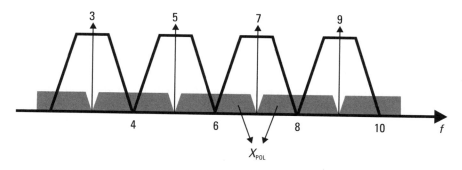

Figure 5.14 Output spectrum in feed terminal when one kind of polarization is tuned (odd transponders).

cross-polar isolation of a receiving antenna, ranging from 20 to 30 dB in the direction of the maximum receiving power (boresight) and then including the effects of any installation misalignments.

Example 5.6

Determine the interference degradation factor ΔI for X_{POL} if considering only the XPI of the receiving antenna with a value of 30 dB. Assume that C/N is 10 dB.

Solution. Using (5.26),

$$\Delta I = -10 \cdot \log\left(1 - 10^{(10-30)/10}\right) = 0.044 \text{ dB}$$

ACI comes from the channels on the same polarization as the wanted received channel but with the immediately higher and lower channel frequencies (not to be confused with the channel number, which may not be contiguous with channel frequency). Because no physical filtering is ever perfect (in particular spacecraft output filters are made with as low a loss as possible, yielding minimum rejection), transmissions in the adjacent channel overlap with the wanted channel to some extent. This is made worse if the adjacent channels are higher in output level. To minimize this problem, spacecraft operators exercise careful frequency, modulation, and level control of all transmissions.

The TI comes from terrestrial sources near the reception system and operating at the same (or very near) frequency as the satellite TV system. This kind of interference is very common in C-band systems where it often stems from terrestrial microwave systems. The TI is usually detected by field measurements. It is very difficult to devise a tractable mathematical model to assess TI in a well-designed satellite TV reception system because of the large number of special cases encountered. Avoidance by better geometry is the primary tool and is much preferred to shielding, which can also be used.

The OI comes from an adjacent satellite that transmits at the same frequency and, perhaps, the same polarization (interferer satellite). The most effective way to reject this kind of interference is by a good selection of the receiving antenna radiation pattern. Figure 5.15 shows a typical situation to evaluate OI without the presence of thermal noise.

The engineering approach to the evaluation of OI comes from several simplifying and conservative assumptions:

- Protection against interference is obtained with the Earth station antenna directivity. For system planning ITU-R antenna sidelobe envelopes are widely used (CCIR-Rep. 391). For small-diameter antennas ($D/\lambda < 100$), it is given by [8]

$$G_R(\theta) = 52 - 25 \log \theta° - 10 \log\left(\frac{D}{\lambda}\right), \text{dBi} \qquad (5.29)$$

for $1° < \theta < 48°$.

- The tuner bandwidth B has an equal or smaller value than the wanted (B_W) and interferer (B_I) transponders.

- The wanted satellite downlink transmission loss has very near the same value as the interferer satellite transmission loss, and L denotes both.

Remembering the basic transmission equation explained in Chapter 3, it is possible to write

$$(C/I)_{OI} = \frac{(\text{EIRP}_W / B_W)(B/L)}{(\text{EIRP}_I / B_I)(B/L)} \cdot \frac{G_R}{G_R(\beta)} \qquad (5.30)$$

where:

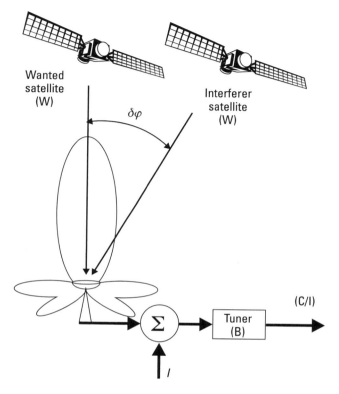

Figure 5.15 OI. $\delta\varphi$ represents the orbital spacing between wanted and interferer satellites. *B* is the tuner bandwidth.

- $(\text{EIRP}_W / B_W)(B / L)$ and $(\text{EIRP}_I / B_I)(B / L)$ represent the RF power at demodulator input for the wanted and interferer carriers, respectively (in watts).

- G_R is the receiving antenna gain (maximum) in the direction of the wanted satellite.

- $G_R(\beta)$ is the receiving antenna gain in the direction of the interferer satellite.

The angle β can be calculated using the geometric model depicted in Figure 5.16, where *T* represents the location of the receiving ground terminal, and S_W and S_I represent the orbital positions of the wanted satellite and the interferer satellite, respectively.

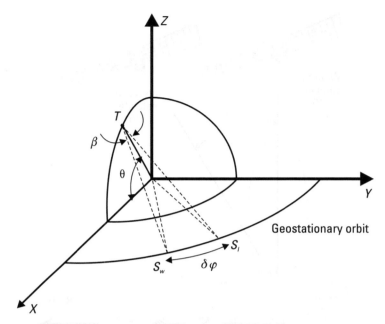

Figure 5.16 Geometric model to calculate β angle. S_W is the wanted satellite and S_I is the interferer satellite.

Let d_S be the distance between the wanted and the interferer satellites, which can be calculated as

$$d_S + (R + H) \cdot \delta\varphi \qquad (5.31)$$

From triangle TS_WS_I,

$$\cos\beta = \frac{d_1^2 + d_2^2 - d_S^2}{2d_1 d_2}$$

where d_1 and d_2 are the distances to satellites S_W and S_I, respectively, from the ground terminal located in T. Assuming that $d_2 \approx d_1$, then

$$\cos\beta = 1 - \frac{d_S^2}{2d_1^2}$$

and

$$\sin(\beta / 2) = \frac{d_S}{2d_1} \tag{5.32}$$

Substituting (5.31) in (5.32) and using (3.10), then

$$\beta = 2 \arcsin \frac{0.494\,\delta\varphi}{\sqrt{1 - 0.296 \cos \Delta\varphi \cos \theta}} \tag{5.33}$$

and it can be further approximated by

$$\beta = \frac{0.988\,\delta\varphi}{\sqrt{1 - 0.296 \cos \Delta\varphi \cos \theta}} \tag{5.34}$$

It is possible to demonstrate that, for relatively small values of $\delta\varphi$ ($\leq 10°$), the $\beta/\delta\varphi$ ratio remains in the range 1.1 to 1.18 [9]. A practical formula is

$$\beta = 1.15\,\delta\varphi \tag{5.35}$$

Substituting (5.1), (5.29), and (5.32) in (5.30), obtain

$$(C/I)_{OI} = \frac{\text{EIRP}_W}{\text{EIRP}_I} \cdot \frac{B_I}{B_W} \cdot \frac{\eta\left(\dfrac{\pi D}{\lambda}\right)^2}{10^{\frac{52 - 25 \log(1.15\delta\varphi) - 10 \log\left(\dfrac{D}{\lambda}\right)}{10}}} \tag{5.36}$$

Assuming equal EIRP and equal transponder bandwidth ($B_W = B_I$), (5.36) can be transformed into

$$\begin{aligned}(C/I)_{OI} &= 10 \log(\eta) - 42.05 \\ &+ 30 \log\left(\frac{D}{\lambda}\right) + 25 \log(1.15 \cdot \delta\varphi), \text{dB}\end{aligned} \tag{5.37}$$

Using a 55% efficiency antenna,

$$(C/I)_{OI} = -43.13 + 30 \log\left(\frac{D}{\lambda}\right) + 25 \log(\delta\varphi), \text{dB} \tag{5.38}$$

Figure 5.17 shows a typical performance of OI for different spacing values between adjacent satellites.

Example 5.7

Let the wanted and interferer satellites depicted in Figure 5.16 have the same EIRP and transponder bandwidth. If the satellites are 2° apart from each other, the operating frequency is 4 GHz (C-band), and the receiver antenna size is 3m, determine the orbital interference.

Solution.   Substituting into (5.38), obtain

$$(C/I)_{OI} = -43.13 + 30\log\left[\frac{3}{0.075}\right] + 25\log(2°) = 12.45, \text{dB}$$

Example 5.8

Using the results of Example 5.7, determine the degradation factor ΔI due to orbital interference if the required C/N is 10 dB (typical in C-band analog systems).

Solution.   Using (5.26),

$$\Delta I = -10\log\left\lfloor 1 - 10^{(10-12.45)/10}\right\rfloor = 3.65\,\text{dB}$$

Example 5.9

Repeat Example 5.7: The interferer satellites are 9° apart from each other, the operating frequency is 12 GHz, and the receiver antenna size is 60 cm.

Solution.   Substituting again in (5.38), obtain

$$(C/I)_{OI} = -43.13 + 30\log\left[\frac{0.60}{0.025}\right] + 25\log(9°) = 22.13\,\text{dB}$$

Example 5.10

Repeat Example 5.9, with a C/N of 6.5 dB (which is typical in a digital system operating in the Ku band).

Solution.   Using (5.26),

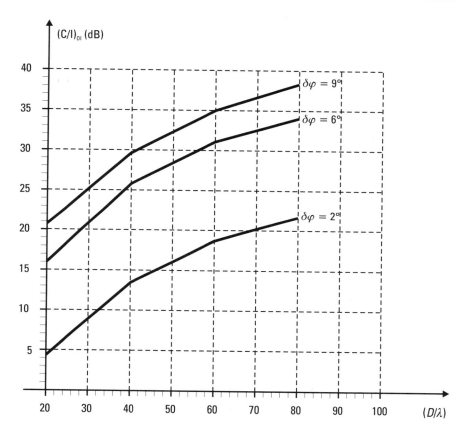

Figure 5.17 Orbital interference performance.

$$\Delta I = -10 \log\left[1 - 10^{(6.5-22.13)/10} \right] = 0.12 \text{ dB}$$

Example 5.11

Satellites Galaxy 5 and Galaxy 9 are located at 125° W and 123° W, respectively, and both operate in the C band (4 GHz). They use the same frequency plan but with inverted alternating polarization. For example, Galaxy 5 operates at 4 GHz with vertical polarization, while Galaxy 9 operates at the same central frequency but with horizontal polarization, and so on. Determine the orbital interference $(C/I)_{OI}$ from Galaxy 9 to Galaxy 5 if the receiving antenna pointed to Galaxy 5 has a 2.5-m diameter, and the cross-polar reference gain pattern is given by the following values (Rec. 652-1):

$$
\begin{array}{rll}
-25\,\text{dB} & \text{for} & 0 \le \theta \le 0.25\,\text{BW}^\circ \\
-\left(30 + 40\log|\theta / \text{BW}^\circ{-}1|\right)\text{dB} & \text{for} & 0.25\,\text{BW}^\circ \le \theta \le 0.44\,\text{BW}^\circ \\
-20\,\text{dB} & \text{for} & 0.44\,\text{BW}^\circ \le \theta \le 1.28\,\text{BW}^\circ \\
-\left(17.3 + 25\log|\theta / \text{BW}^\circ|\right)\text{dB} & \text{for} & 1.28\,\text{BW}^\circ \le \theta \le 3.22\,\text{BW}^\circ
\end{array}
$$

relative to the gain (maximum) of the main beam. Assume the same EIRP values for both satellites.

Solution. The receiving antenna beamwidth is

$$
\text{BW}^\circ = 70 \cdot \frac{0.075}{2.5} = 2.1^\circ
$$

and the angle β is

$$
\beta = 1.15 \cdot 2^\circ = 2.3^\circ
$$

The interval of interest for the cross-polar reference pattern is the third one, given by the following range of values:

$$
0.924^\circ \le \theta \le 2.688^\circ
$$

and the corresponding cross-polar isolation is 20 dB.

From (5.38), the OI is 10.08 dB for $\delta\varphi = 2^\circ$. Adding the cross-polar isolation in the direction of interferer satellite Galaxy 9, the total orbital interference rejection is 30.08 dB. The interference degradation factor assuming a C/N of 10 dB is

$$
\Delta I = -10\log\left[1 - 10^{(10-30.08)/10}\right] = 0.043\,\text{dB}
$$

and it may be neglected for practical applications.

In conclusion, it is possible to say the following.

- ACI and X_{POL} can be neglected in practical applications for the downlink design as observed in Example 5.6. The ACI rejection should be better in DBS Ku-band systems than in C-band systems. These interference tolerance limits are mainly taken into account in the planning stage of satellite TV systems and should not be an important degrading factor for reception quality.

- TI mainly affects C-band systems where terrestrial microwave systems can operate at the same frequencies. Some geometrical placement and shielding methods can be used to avoid this kind of interference. Ku-band systems in BSS are not affected by terrestrial interference, because this frequency band should not be used in any other radiocommunication service.

5.6.3 Implementation Loss

Receiver performance is characterized in terms of an implementation margin with respect to a perfect receiver model. This leads to an implementation loss for acceptable service, which should be used in the link budget analysis.

The main sources contributing to the implementation loss can be listed as follows:

- An important element in the implementation loss is the LNB, which, in addition to thermal noise, is a source of phase noise. Phase noise not only contributes to the implementation loss but may also be the cause of the rate increase of cycle slipping in the carrier synchronization circuits.

  The signal degradation due to phase noise in a LNB may be modeled as an additional noise source with a respective C/N_ϕ (ϕ: phase noise). Let C/N represent the overall C/N in the QPSK demodulator input and $(C/N)_{th}$ represent the thermal noise contribution. Because the phase noise power is small and can be considered as an equivalent noncoherent Gaussian noise source, it is possible to write

$$\left(C/N\right)^{-1} = \left(C/N\right)_{th}^{-1} + \left(C/N\right)_{\phi}^{-1} \qquad (5.39)$$

Using some algebraic manipulations,

$$\left(C/N\right)_{th} = \left(C/N\right) \cdot L_{imp} \qquad (5.40)$$

where the impairment loss is defined as

$$L_{imp} = \left[1 - \frac{(C/N)}{(C/N)_\phi} \right]^{-1} \tag{5.41}$$

C/N_ϕ is the reciprocal value of the integral (in the frequency range of interest) of the two-sided phase noise power spectral density (dBc/Hz) at the QPSK demodulator input. For example, if $C/N = 8$ dB (typical in digital DVB-S receivers), and $C/N_\phi = 25$ dB, then L_{imp} = 0.087 dB.

- Standing waves generated on the first IF frequency band as a result of LNBF output and IRD tuner input mismatching. If ρ represents the magnitude of the reflection coefficient ($0 \le \rho \le 1$), then the impairment loss L_{mis} due to mismatching in a connection point is

$$L_{mis} = -10 \log\left(1 - \rho^2\right), dB \tag{5.42}$$

where

$$\rho = \frac{VSWR - 1}{VSWR + 1} \tag{5.43}$$

and VSWR is the voltage standing wave ratio. For example, if VSWR is 1.3:1 in the LNBF output, then the impairment loss due to mismatching in that point is 0.074 dB. Usually, the VSWR value is not fairly constant over the entire frequency range of operation so this variation can also contribute to the implementation loss.

- The Viterbi decoder: The number of input quantization levels used as well as the trace-back memory depth will influence Viterbi decoder performance. Practical implementations will normally fall within 0.2 to 0.3 dB of the performance of an ideal decoder and may be incorporated into the coding gain.

- Receive channel filtering: The implementation and complexity of the receive half-Nyquist filtering have to be chosen as a compromise between receiver cost and performance. Values on the order of 1 to 1.5 dB are commonly found in practice [10].

- The satellite, although outside the receiver, should be considered as another source contributing to implementation loss. The satellite has a nonlinear amplifier (usually a TWTA) with bandpass filters at its input (IMUX) and output (OMUX) multiplexers. The TWTA (assumed saturated) contributes an implementation loss that is relatively flat with frequency and that is roughly independent of the precise modulated carrier bandwidth. The IMUX and OMUX filters contribute an implementation loss that varies depending on the spectral occupancy (the wider the modulated carrier spectrum, the greater the degradation). This is due not only to the bandwidth limitation but also to the group delay irregularity at the band edges. Typical degradation values in (E_b/N_0) at the input of a QPSK demodulator are less than 1 dB for $(B_T/R_S) \geq 1.10$.

- Antenna pointing losses due to an antenna's misalignment as a consequence of wind or other environmental conditions and/or satellite station-keeping accuracy around its nominal orbital position: A typical value for small antennas (< 1m) is 0.5 dB.

The implementation losses can be included in the term ΣL ("other losses") in the margin equation.

5.6.4 Downlink Design

The goal in choosing components for a satellite reception system is to produce a signal with enough power to properly drive a television receiver, stereo, or computer terminal. Although standards for acceptable signal quality vary from those necessary for SMATVs to those adequate for a home satellite reception system, in all cases the signal entering a satellite receiver should have a C/N ratio sufficiently high so that an analog or digital receiver can recover the baseband signal of video, audio, or data.

The two choices that are by far the most important in designing a satellite TV reception system are the receiver antenna gain and the LNB noise temperature. Taking into consideration the definition of G/T, it is possible to write

$$G_R = 10^{(G/T)/10} \cdot \left(T_A + T_{LNB}\right) \tag{5.44}$$

where G/T is expressed in dB/K and G_R is the receiver antenna gain (not in decibels). Using (5.1), it is possible to write

$$D = \frac{\lambda}{\pi} \sqrt{\frac{10^{(G/T)/10} \cdot \left(T_A + T_{LNB}\right)}{\eta}}, \text{m} \tag{5.45}$$

The diameter thus obtained takes into consideration only the noise limitation of the satellite downlink. If it is necessary to protect against an adjacent interferer satellite (OI), a new diameter must be chosen to satisfy a specific required value for the C/I as it is expressed in (5.38). Let us call D_1 the diameter obtained by (5.45), and D_2 that obtained by means of (5.38). Then, the diameter selection must obey

$$D = \max(D_1, D_2) \qquad (5.46)$$

In almost all practical situations, the satellite downlink is limited by noise. In C-band systems, where adjacent satellites may be 2° or 3° apart from each other, it is possible that OI and/or TI can also limit that satellite downlink. The flowchart in Figure 5.18 establishes a general guideline to the evaluation of antenna diameter.

In order to estimate C/I in the interference branch shown in Figure 5.18, it is possible to carry out the following procedure:

1. Set ΔI at some value—for instance, $\Delta I \leq 1$ dB.
2. Estimate C/I using the inequality

$$C/I \geq C/N - 10\log\left(1 - 10^{-\Delta I/10}\right), dB \qquad (5.47)$$

where C/N is in the range of 10–14 dB for analog systems and 6–8 dB for digital systems.

If the interfering carrier has an orthogonal polarization in respect to the wanted carrier, then, it is necessary to add the XPI value (decibels) to C/I.

Example 5.12

Determine the minimum antenna diameter for a digital satellite TV reception system located in Berlin (52.6° N, 12.4° E) receiving linear polarized signals from ASTRA 1F (19.2° E) at 12.1–12.5 GHz and 52 dBW. The reception system is compliant with the DVB-S standard. The symbol rate is 27.5 Msymb/second and the convolutional code rate used is 3/4. The transponder bandwidth is 33 MHz, the uplink contribution noise is 0.2 dB, and the antenna noise temperature (clear-sky conditions) is 35K and 65% of efficiency. The LNB noise figure is 1.1 dB, the antenna pointing losses are 0.5 dB, the external interference is 1 dB, the feed loss is 0.2 dB, and the satellite

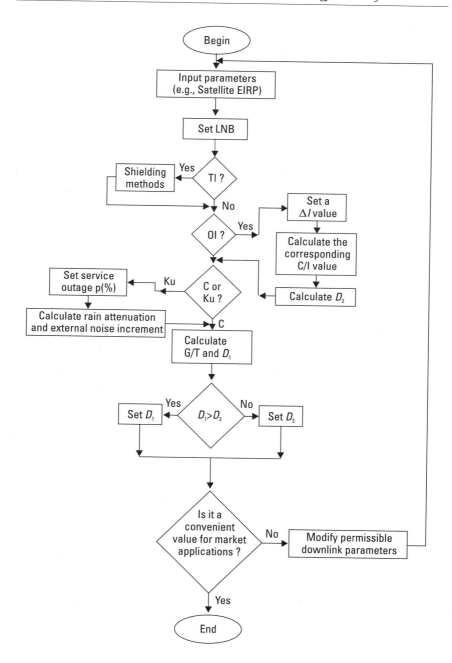

Figure 5.18 Guideline for downlink analysis.

and receiver degradation is 1.8 dB. The required service outage is, on average, 0.1% average year.

Solution. The elevation angle EL° is

$$EL° = \arctan\left[\frac{\cos(6.8°)\cos(52.6°) - 0.1513}{\sqrt{1 - \cos^2(6.8)\cos^2(52.6°)}}\right] = 29.52° \cong 30°$$

Berlin is located in zone E for rain calculations using the ITU-R method. Then, $R_{0.01\%}$ = 22 mm/hr. See Table 5.2 for the results with H and V polarization and f = 12.5 GHz (worst case). The worst case occurs with horizontal polarization.

The specific rain attenuation for 0.01% service outage is

$$\gamma_{R0.01\%} = 0.028 \cdot (22)^{1.207} = 1.168 \text{ dB / km}$$

Assuming that the receiving site is 100m above average sea level, the effective slant-path length L_e through rain can be calculated as follows:

$$L_e = L_S \cdot r_{0.01}$$

$$L_S = \frac{5 - 0.075 \cdot (52.6 - 23) - 0.1}{\sin(30°)} = 5.36 \text{ km}$$

$$L_0 = 35 \cdot e^{-0.015 \cdot 22} = 25.16 \text{ km}$$

$$r_{0.01} = \frac{1}{1 + \dfrac{5.36}{25.16}\cos(30°)} = 0.844$$

and, finally

$$L_e = 5.36 \cdot 0.844 = 4.525 \text{ km}$$

The rain attenuation for 0.01% service outage (the average amount per year) is

Table 5.2
Solution for Example 5.12

Parameter	Polarization	
	H	**V**
k	0.028	0.0192
α	1.207	1.193

$$A_R(0.01\%) = 1.168 \cdot 4.525 = 5.2825 \text{ km}$$

For 0.1% service outage (the average amount per year), the rain attenuation is

$$A_R(0.1\%) = 5.2852 \cdot 0.12 \cdot (0.1)^{-[0.546 + 0.043 \cdot \log(0.1)]} = 2.01945 \text{ dB} \cong 2.02 \text{ dB}$$

The external noise increment ΔT due to rain is (3.54)

$$\Delta T = 10 \log \left[1 + \frac{240 \cdot (1 - 10^{-0.202})}{50 + 290 \cdot (10^{0.11} - 1)} \right] = 2.22 \text{ dB}$$

The required rain margin M_R is

$$M_R \geq A_R + \Delta T = 2.02 + 2.22 = 4.24 \text{ dB}$$

so then, $M_R = 4.3$ dB.

The free-space loss is

$$L_b = 185 + 20 \log 12.5 + 10 \log[1 - 0.296 \cos(6.8°) \cos(52.6°)] = 206.08 \text{ dB}$$

and the additional loss is $\Sigma L = 0.5 + 0.2 + 1.8 = 2.5$ dB.

The roll-off factor α is calculated using (4.6). Thus,

$$\alpha = \frac{33}{27.5} - 1 = 0.2$$

Then, the MPEG-2 TS information bit rate R_b can now be calculated as

$$R_b = \left(\frac{2}{1+0.2}\right) \cdot \left(\frac{188}{204}\right) \cdot \left(\frac{3}{4}\right) \cdot 33 = 38.01 \text{ Mbps}$$

Using decibels per bits per second,

$$R_b = 10 \log 38.01 \cdot 10^6 = 75.79 \text{ dB.bps}$$

The coding gain G_C for a 3/4-convolutional code rate is 8.0 dB. The G/T value is then

$$G/T = 43 - 52 + 206.08 + 2.5 + 75.79$$
$$+ 13.5 + 0.2 + 1 - 8.0 - 228.6 = 14.77 \text{ dB/K}$$

The LNB noise temperature can be calculated using the well-known conversion formula between noise temperature and noise figure; that is,

$$T_{LNB} = 290 \cdot \left(10^{0.11} - 1\right) = 83.59\text{K}$$

and the system noise temperature is

$$T_A + T_{LNB} = 10^{-0.02} \cdot 35 + 290\left(1 - 10^{-0.02}\right) + 83.59 = 130.09\text{K}$$

Finally, we obtain

$$D = \frac{0.024}{\pi} \cdot \sqrt{\frac{10^{14.77/10} \cdot 130.09}{0.65}} = 59.2 \text{ cm}$$

and a size of 60 cm would be adopted for practical purposes.

Example 5.13

Select a suitable antenna diameter to receive analog TV signals in a location at 22° N, 80° W from Galaxy V (125° W). The LNB noise temperature is 25K, and the receiver (IRD) bandwidth is 30 MHz (at the second IF). Assume 0.5 dB for atmospheric noise and 0.5 dB for uplink noise contribution. The EIRP is 35 dBW, and the satellite is operating at 4 GHz (C band). The TV standard is NTSC-M, and it is necessary to provide a quality grade of four (UIT-R five-point scale). The peak-to-peak frequency deviation is 25 MHz/V, and the transponder bandwidth is 36 MHz.

Solution.   Using (4.1), for $Q = 4$

$$S / N = 23 - 4 + 1.1 \cdot (4)^2 = 36.6 \, \text{dB}$$

The C/N value can be calculated using (4.2) without considering the weighting factor *pw*. Then,

$$C / N = 36.6 - 10 \log \left[\frac{3}{2} \cdot \left(\frac{25}{4.2} \right)^2 \cdot \left(\frac{36}{4.2} \right) \right] = 10 \, \text{dB}$$

Let us calculate the following parameters:

- Free-space loss:

$$L_b = 185 + 20 \log 4$$
$$+ 10 \log \left[1 - 0.296 \cdot \cos(-45°) \cdot \cos(22°) \right] = 196.1 \, \text{dB}$$

- Receiver bandwidth in decibels per bits per second:

$$R_b = 10 \log \left(30 \cdot 10^6 \right) = 74.8 \, \text{dB.bps}$$

- Elevation angle:

$$EL° = \arctan \left[\frac{\cos(-45°) \cdot \cos(22°) - 0.1513}{\sqrt{1 - \cos^2(-45°) \cdot \cos^2(22°)}} \right] = 33.7$$

- Using (5.7), the antenna noise temperature is

$$T_A = \frac{77}{D} + \frac{454}{33.7} = \frac{77}{D} + 13.4 \text{K}$$

- Because rain has practically no effect in the C band, the margin can be selected as 1 dB.

The G/T can be calculated using (3.57) backwards:

$$G / T = 1 - 35 + 10 + 196.1 + 0.5 + 74.8 + 0.5 - 228.6 = 19.3 \text{ dB/K}$$

Assuming 55% for the antenna efficiency, it is possible to write the following equation:

$$0.55 \cdot \left(\frac{\pi \cdot D}{0.075} \right)^2 = 10^{1.93} \cdot \left(\frac{77}{D} + 13.47 + 25 \right)$$

which can be transformed to the cubic equation

$$D^3 - 3.4D - 6.8 = 0$$

and $D = 2.5$m.

The receiving antenna diameter calculated in the way just shown should warrant a minimum value of (E_b/N_0) at the QPSK demodulator input for a specific *quality of service* (QoS) (a determined BER for a probability service outage). The receiving antenna diameter (or gain) also warrants a minimum power level at the IRD input for the expected performance of electronic equipment. Let P_R (in decibels relative to 1 mW) be the power level at the receiving antenna feed output, and S_R (also in decibels relative to 1 mW) be the minimum power level at the IRD input (sensitivity). Referring to Figure 5.19, it is possible to write

$$S_R(\text{dBm}) = P_R(\text{dBm}) + G_{\text{LNB}}(\text{dB}) - L_X(\text{dB}) \qquad (5.48)$$

where G_{LNB} is the LNB's conversion gain and L_X represents the coaxial loss between the LNB output and IRD input at the higher frequency in the first IF frequency range. L_X can also represent the net loss of a SMATV system for community reception. Using basic concepts in link budget analysis (see Chapter 3), it is possible to write the following equation:

$$G_R(\text{dBi}) = P_R(\text{dBm}) - \text{EIRP}(\text{dBW})$$
$$+ L_b(\text{dB}) + \Sigma L(\text{dB}) + A_R(\text{dB}) - 30 \qquad (5.49)$$

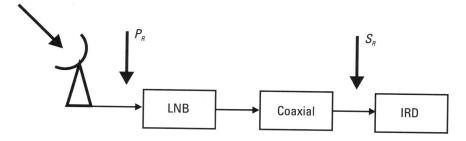

Figure 5.19 IRD sensitivity analysis.

where A_R (in decibels) represents the rain attenuation, and the term 30 is used for unit conversion between decibels relative to 1W and decibels relative to 1 mW. If the receiving antenna gain calculated from (5.48) and (5.49) is higher than that obtained from (5.46), then it is necessary to use a new antenna diameter given by

$$D_{new}(m) = D(m) \cdot 10^{\Delta G(dB)/20} \qquad (5.50)$$

where ΔG (in decibels) represents the difference (> 0) in antenna gains, in decibels. The receiving antenna diameter can be increased until practical or until economical limits are reached. Otherwise, line amplifiers should be used to boost the power level at the IRD input.

Example 5.14

Determine the maximum loss allowable in a SMATV-IF distribution system (L_x) if the receiving system has the following parameters:

- Antenna diameter: 60 cm;
- Antenna efficiency: 60%;
- Operating frequency: 12.5 GHz;
- IRD sensitivity: –60 dBm;
- LNB conversion gain: 60 dB;
- Satellite EIRP: 51 dBW;
- Additional loss: 3.5 dB;
- Rain attenuation: 8 dB (99% service availability in the average year).

Solution. The antenna gain is

$$G_R = 10 \log \left[0.6 \left(\frac{\pi \cdot 0.60}{0.024} \right)^2 \right] = 35.68 \text{ dBi}$$

The maximum loss (L_x) of the SMATV-IF system can be calculated by combining (5.48) and (5.49). Then,

$$L_x = 35.68 + 60 + 60 + 51 - 205.4 - 3.5 - 8 + 30 = 19.78 \text{ dB}$$

5.7 DVB-S for Wireless Internet Access

Internet services, based on *transfer control protocol/Internet protocol* (TCP/IP), were initially developed to support research organizations and commercial enterprises. In developed countries, the focus has shifted from these organizations to the people.

The Internet is becoming popular and taking a more vital role in how we work, interact, and even enjoy our free time. With this increasing popularity, home access to the Internet has increased significantly. Most access circuits available to end users employ copper cables, and most users access the Internet via a dial-up model and a *public switched telephone network* (PSTN). However, the data transmission speed often poses a problem. Large files could take hours to transmit, tying up important computer resources. Therefore, high-speed network access circuits are desired.

In the present multimedia communication environment, the predominant communication style is for user terminals to access terminals to obtain a lot of multimedia information (Figure 5.20).

Some terrestrial technologies take into consideration the asymmetric traffic generated in Internet access. These are listed in Table 5.3.

ADSL and cable modems (used in conjunction with CATV systems) can achieve higher data rates than those shown in Table 5.3. For example, *asymmetric digital subscriber loop* (ADSL) has data rates ranging from 1.5 to 9 Mbps (high-capacity channel) while cable modems can achieve up to 27 Mbps.

Nonetheless, satellite systems have unique features that wireline terrestrial systems do not have. These include the following:

- Broadcast capability (point-to-multipoint communications);

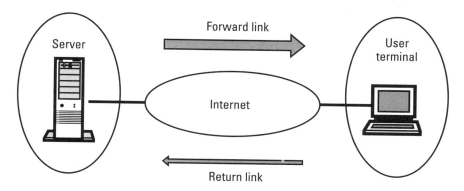

Figure 5.20 Communication style for multimedia communications.

- Wide coverage area;
- Easy installation of Earth stations;
- Quick establishment of communication links.

So, satellite systems can achieve delivery of Internet information to end users by broadcasting at high data rates up to 30–45 Mbps using the DVB-S standard and technologies over a typical transponder bandwidth. However, to allow interactivity, a return channel from the end-user terminal to the end Internet service provider is required. If this return channel were established using a satellite link, the cost for user stations would be excessive.

Table 5.3

Data Rate Comparison for Users' Internet Access Using Different Terrestrial Technologies

Kind of Channel	Data Rate	Kind of Internet Access
14.4 modem	14.4 Kbps	Basic analog Internet access method
28.2 modem	28.2 Kbps	Basic analog Internet access method
ISDN_single B channel	64 Kbps	Expensive digital connection to the Internet
ISDN_dual B channel	128 Kbps	Considered the fastest Internet access for residential and business

Consequently, the high-speed satellite link is reserved for the downstream flux, and a low-speed terrestrial link, using PSTN or dedicated circuits, is used as the return channel. The basic architecture approach for this kind of Internet access is shown in Figure 5.21. IP packets must be embedded in 188-byte MPEG-2 TS packets for satellite transmission and reception over the DVB-S standard.

The *PC interface* (PCI) performs the following functions:

- Downloading and rebuilding the IP packets embedded in the 188-MPEG-2 TS. It is possible, when using high-speed terrestrial circuits like *asynchronous transfer mode* (ATM) over *synchronous digital hierarchy/synchronous optical network* (SDH/SONET), that the 188-byte MPEG-2 TS packet is first downloaded to 54-byte ATM cells and then to IP packets.

- Filtering out those IP packets that are not addressed to the user terminal. It should be assumed that the satellite downlink receives IP packets for different user terminals that were time-multiplexed at the satellite uplink transmission site.

Example 5.15

Determine the cost per month and per user for a satellite broadcast channel using the standard DVB-S, for Internet access, with the following parameters:

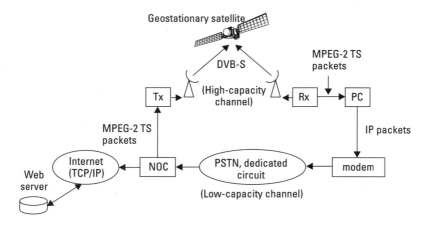

Figure 5.21 Satellite system using DVB-S for Internet access (NOC: network operating center).

- A QPSK modem with a roll-off of 0.2 in the shaping filters;
- A convolutional code rate of 3/4;
- A 24-MHz transponder bandwidth.

The traffic estimate is 2 Mb/user/hour, and such a channel can be operated at a loading factor of 60% without a significant impact on the QoS to individual users. The 24-MHz transponder cost is $200,000 per month.

Solution. Using (4.16), the satellite downlink capacity can be calculated as

$$R_b = \frac{1.84}{(1+0.2)} \cdot (3/4) \cdot 24 = 27.65 \text{ Mbps}$$

A fully loaded 27.65-Mbps DVB-S channel during this 10-hour business day has room for

$$\frac{60 \cdot 60 \cdot 27.65 \cdot 10^6}{2 \cdot 8 \cdot 10^6} = 6,221.25 \text{ (users)}$$

and, for a loading factor of 60%, this gives us a capacity of about 37,327 users. Thus, the cost in transponder time for the broadcast satellite channel is about $5.36/month per user.

A typical product using this kind of satellite technology for Internet access is DirecPC, which includes a 24-inch dish antenna and a PCI adapter. Another product is DirecDuo. This system eliminates the need for two dishes. It is an elliptically shaped parabola that picks up the TV satellite and the Internet satellite. It has a three-way LNB, so it is possible to connect to the Internet with a cable to a computer and add two or more TV receivers. The cost of both the end users' ground terminal and the monthly cost per user of the shared satellite broadcast channel is moderate because it must be competitive with terrestrial means via a cable modem or ADSL. One kind of service supported on DirecPC or DirecDuo is Turbo Internet, a two-way interactive service where subscribers use a Web browser (Explorer or Navigator) to make their information requests over a modem linked to a telephone line. DirecPC (or DirecDuo) receives the information request at its NOC, obtains the data from its IP gateway to the Web, and then relays the data over the satellite link to the end user. Another service is Turbo Webcast, where DirecPC subscribers can select from a list of the most popular sites on the Internet and have those sites delivered automatically to their hard drives over high-speed satellite links. Because the information is delivered directly to their PC hard drives, access to those sites is instantaneous and the subscribers' telephone lines remain free for domestic or business use. Turbo Newscast is another service implemented on DirecPC. This service allows DirecPC subscribers to select thousands of Usenet newsgroups and have the contents of those newsgroups delivered in a way similar to the delivery method of Turbo Webcast.

References

[1]   Baylin, F., *1998/2000 World Satellite Yearly*, Boulder, CO: Baylin Publications, 1998, p. 18.

[2]   Maral, G., and M. Bousquet, *Satellite Communication Systems*, New York: John Wiley and Sons, 1993, p. 336.

[3]   Reston, W., *Communication Services Via Satellite*, New York: John Wiley and Sons, 1985, pp. 294–298.

[4]   Mueller, K., "A Low-Cost DVB-Compliant Viterbi and Reed-Solomon Decoder," *IEEE Trans. on Consumer Electronics*, Vol. 43, No. 3, August 1997, pp. 448–450.

[5]   Tomasz, M., "Direct-Conversion Receiver IC Simplifies DBS Set-Top Boxes," *Microwaves & RF*, March 1997, pp. 169–172.

[6]   Ritcher, K., "Zero IF Satellite Tuners for DBS," *IEEE Trans. on Consumer Electronics*, Vol. 44, No. 4, November 1998, pp. 1367–1370.

[7]   Muschallik, C., "System Considerations on SCPC for Digital Satellite Receivers with Direct Conversion," *IEEE Trans. on Consumer Electronics*, Vol. 45, No. 3, August 1999, pp. 965–969.

[8]   Pritchard, W., and M. Ogata, "Satellite Direct Broadcast," *Proc. IEEE*, Vol. 78, No. 7, July 1990, pp. 1116–1140.

[9]   Maral, G., *VSAT Networks*, New York: John Wiley and Sons, 1995, p. 211.

[10]   Stephenson, D. J., *Guide to Satellite TV*, Oxford: Newnes, 1997, p. 162.

6

Satellite DAB Systems

6.1 Introduction

Three major allocations involving the broadcasting and broadcasting-satellite services were discussed at the 1992 WARC (WARC'92):

1. Expansion of frequencies for conventional high-frequency (short-wave) broadcasting;
2. New frequencies for both satellite and terrestrial digital audio (radio) broadcasting to handheld and automobile receivers;
3. New frequencies for HDTV broadcasting from satellites.

Of these three broadcasting issues, the proposal for DAB is perhaps the most exciting. ITU-R has allocated the frequency band L (1.5 GHz)—the S band for regions that use the L band for other services, such as North America—for this kind of service. Digital coding and modulation techniques that are already developed and demonstrated permit compact disc (CD)-quality stereo signals to be broadcast from either satellite or terrestrial transmitters to portable and vehicular receivers without the fading typical of car-radio reception near the edge of broadcast service areas.

The role of satellites in mobile radio broadcasting is increasingly assured. Automobile industry giants were quick to realize this. General Motors and Ford are involved in the development of two satellite radio broadcasting systems: XM Satellite Radio (with Hughes as the primary

161

satellite contractor) and Sirius—formerly Satellite CD Radio (with Loral as the primary satellite contractor).

The world is entering a new age of satellite radio broadcasting, with new systems offering wide coverage areas, high-quality sound, user-friendliness, and both portable and mobile reception. Conventional radio broadcasting services, like AM and FM, cannot give equivalent reach and performance. The digital format also makes it possible to offer complementary services, such as multimedia and Internet data transmission.

Like TV, radio is turning to digital broadcasting for several reasons:

- Audio signal compression has improved to such an extent that digital signals now occupy less bandwidth than analog signals of the same quality and much less bandwidth than with the CD standard.

- Signals are more resistant to external interference, thanks to the possibilities provided by digital processing.

- Digital audio signals can be multiplexed with any other signals (e.g., images and text files), making it possible to enrich the content.

- Digital signal processing is cheaper than analog signal processing, thanks to the use of highly integrated silicon components.

Audio broadcasting is a challenging broadcasting technology because vehicular and portable receivers provide almost no antenna gain for boosting the power level at the receiver's input. Moreover, foliage absorbs the signals and buildings reflect them, causing the signal level to vary rapidly in time, or fade.

Satellites offer a good choice for digital radio broadcasting but are only competitive with terrestrial broadcasting if a number of requirements are met. These requirements are related to the following factors:

- The broadcaster's access to the satellite, which must be as direct and simple as possible;

- The end-user satellite access, which must be uninterrupted at all times and in all circumstances, whatever the receiving conditions (portable and/or mobile). The satellite transmitter must be sufficiently powerful (tens to hundreds of watts per stereo channel, depending on downlink frequency and antenna beam coverage).

6.2 Satellite Downlink Performance for Land Mobile Receivers Using Geostationary Satellites

For some years many activities aimed at the introduction of land mobile satellite communication services have been undertaken by different organizations all over the world. For example, the MSAT-X program of NASA (United States); the MSAT program of DOC (Canada); the Mobilest program of AUSSAT (Australia); and the PRODAT program of the European Space Agency (ESA).

Satellite communications with land mobile terminals suffer from strong variations of the received signal shadowing and multipath fading. Shadowing of the satellite signal by obstacles in the propagation path (e.g., buildings, bridges, and trees) results in attenuation over the total bandwidth (i.e., flat fading). This attenuation increases with carrier frequency, that is, it is more noticeable at the L band than at UHF. For a low satellite elevation angle the shadowed areas are longer than for high elevation angles. Therefore, the satellite elevation angle is of paramount importance for keeping the required satellite link margins to a reasonable level for a specified QoS.

Most satellite radio broadcasting systems are going to use geostationary satellites to provide high elevation angles to mobile land terminals. Nonetheless, ESA has proposed satellites in *highly elliptical orbits* (HEOs) to beam lower power signals to countries above 40° in northern altitudes, and the Sirius (formerly Satellite CD Radio) constellation has been changed from geostationary to elliptical inclined orbits.

Multipath fading occurs because the satellite signal may be received not only via the direct path but also after being reflected from objects in the surroundings (Figure 6.1). The radio propagation environment is a very complex one for this kind of system and for modeling, but there has been extensive research on this kind of system. Therefore, for all types of land mobile systems, the communication link between the satellite and the mobile receiver is the most critical part of the overall transmission link and limits the performance of the total system. The satellite downlink for mobile reception is shown, very approximately, in Figure 6.2.

Remembering the transmission equation (3.1),

$$C / N = \frac{P_T G_T G_R}{k L_b L_a TB} \qquad (6.1)$$

The basic geometric relation among solid angle Ψ, distance d, and area S on a satellite-concentric spherical surface is

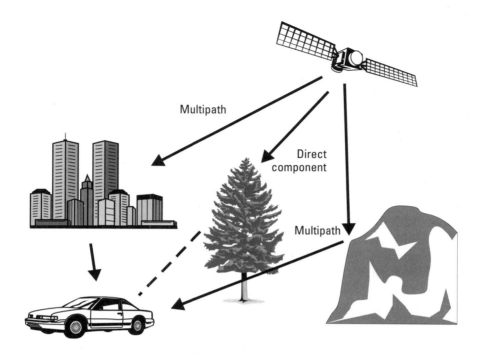

Figure 6.1   Mobile-land satellite downlink: shadowing and multipath.

$$\Psi = \frac{S}{d^2} \text{ steradians} \qquad (6.2)$$

If it is assumed that the radiated power is concentrated in the main beam, the satellite antenna gain G_T is inversely proportional to the beam's solid angle

$$G_T = \frac{K_p}{\Psi} = \frac{K_p d^2}{S} \qquad (6.3)$$

where K_p is a constant of proportionality. Remembering that the free-space loss is (3.3)

$$L_b = \left(\frac{4\pi d}{\lambda}\right)^2 \qquad (6.4)$$

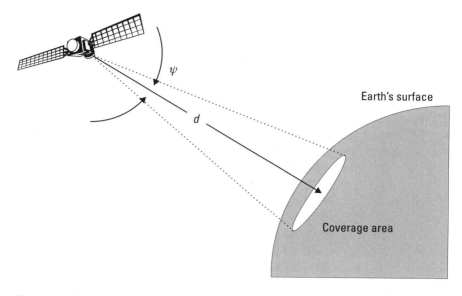

Figure 6.2 Satellite downlink for land mobile receivers.

and assuming that the receiving antenna in the mobile terminal is an isotropic source (G_R = constant); then, substituting (6.4) and (6.3) in (6.1) and making some algebraic manipulations, we obtain

$$C/N = \left[\frac{K_p \cdot G_R}{k \cdot S \cdot (4\pi)^2} \right] \cdot \frac{P_T \lambda^2}{L_a T \, B} \qquad (6.5)$$

where the term in brackets has a constant value. From (6.5), it is possible to reach the following conclusions:

- As the carrier frequency decreases its value, the value of C/N increases.

- Because of the geometry of geostationary systems with respect to the coverage areas, the C/N is effectively not dependent on the distance between the satellite transmitter and the mobile receiver terminal. This compares well with terrestrial radio systems, which must cope with significant near-far effects.

Example 6.1

For system requirements, a 30-dBi parabolic antenna is needed onboard the spacecraft. Determine the aperture diameter for the following frequencies: 500 MHz and 1.5 GHz. Assume that the antenna's efficiency is 55%.

Solution. Rearranging (5.1), the parabolic aperture D can be calculated by

$$D = \frac{\lambda}{\pi} \sqrt{\frac{10^{G(\text{dBi})/10}}{\eta}}$$

Then

    1. f = 500 MHz

$$\lambda = \frac{300}{500} = 0.6\text{m}$$

$$D = \frac{0.6}{\pi} = \sqrt{\frac{10^3}{0.55}} = 8.14\text{m}$$

    2. f = 1.5 GHz

$$\lambda = \frac{0.3}{1.5} = 0.2\text{m}$$

$$D = \frac{0.2}{\pi} = \sqrt{\frac{10^3}{0.55}} = 2.7\text{m}$$

Example 6.2

Determine the power increase for a DAB satellite if the frequency increment is from 1.5 GHz to 3 GHz and all other parameters remain constant.

Solution. Using (6.5), it is possible to write

$$P_T\left(3\,\text{GHz}\right)=\left(\frac{3}{1.5}\right)^2 P_T\left(1.5\,\text{GHz}\right)=4P_T\left(1.5\,\text{GHz}\right)$$

This represents 6 dB more power at 3 GHz than at 1.5 GHz.

Example 6.3

A geostationary DAB satellite located at 95° W is used to cover a circular zone, which has a diameter of 500 km and a center approximately located at 22° N, 80° W. Compare the power requirements with an LEO satellite that has a constant height of 1,200 km. Assume the following parameters:

1. The bandwidth is measured between −3-dB points of the antenna radiation pattern.
2. The antenna efficiency is 55%.
3. The EIRP has a value of 50 dBW.

Solution. The distance between the geostationary satellite and the center of the coverage area can be calculated by

$$d = 4.26\cdot 10^4\sqrt{1-0.296\cos\left(-15^\circ\right)\cos\left(22^\circ\right)}=36{,}519.5\,\text{km}$$

Assuming a right triangle, it is possible to write

$$\tan\left(\frac{\text{BW}^\circ}{2}\right)=\frac{500}{36{,}519.5}=0.01369$$

and

$$\text{BW}^\circ = 1.57^\circ \approx 1.6$$

Combining (5.1) and (5.2) brought us the very useful formula (5.4), which may be written as

$$G(\text{dBi})=44.2-20\log\text{BW}^\circ$$

where the 55% efficiency value has been used. Then,

$$G_T = 44.2 - 20\log(1.6°) = 40.1\,\text{dBi}$$

and

$$P_T = 50 - 40.1 = 9.9\,\text{dBW} \approx 10\,\text{dBW}$$

Repeating the computation for the LEO satellite

$$\tan\left(\frac{\text{BW}°}{2}\right) = \frac{500}{1,200} = 0.417$$

and BW° = 45°.

The corresponding gain for the transmitting antenna onboard the LEO satellite is

$$G_T = 44.2 - 20\log 45° = 11.1\,\text{dBi}$$

and the output power P_T is

$$P_T = 50 - 11.1 = 38.9\,\text{dBW}$$

that is, $P_T = 7.8$ kW.

6.3 Radio Propagation Characteristics in the L-Band Satellite Downlink for Land Mobile Receivers

The satellite radio propagation environment for DAB systems can be divided into three subenvironments: indoor, rural-suburban mobile, and urban mobile [1]. Indoor propagation effects are largely determined by construction material. Nonmetallic materials afford direct, albeit attenuated, penetration of the satellite signal with a minimum of multipath signal scattering. Signal penetration into structures using significant metallic materials is often indirect, through openings such as doors and windows and propagation will involve significant multipath components. Thus, the delay spread in many situations is on the order of tens of nanoseconds, resulting in relatively flat fading. As a result, frequency diversity techniques such as OFDM and *code division multiple access* (CDMA) or equalization techniques do not achieve their intended performance enhancement. Antenna diversity, directivity, and placement are key mitigation techniques for the indoor environment. The simplest mitigation technique is the listener's effort to place the radio or its antenna in a good signal location. This approach may be somewhat hampered by the proximity effect of the listener's body as shadowing.

In the rural-suburban environment with elevation angles greater than 20° (typical when using geostationary satellites), multipath components from the satellite signal are 15–20 dB below the line-of-sight (direct component) signal level, because the receiving antenna radiation pattern of the mobile terminal is often directive in elevation, discriminating low-elevation signal scattering and multipath reflections from the ground (Figure 6.3).

Thus, shadowing is the dominant signal impairment. In addition, fading effects are again found to be relatively flat for a large fading margin, and frequency diversity or spread spectrum techniques are not effective.

Experimental tests have shown that the fade attenuation due to shadowing (A_F) depends on the elevation angle (EL°) and service outage probability [p (%)]. The following empirical expression (taken from an optimal curve fit) to estimate an approximated value of A_F in the L band (1.5 GHz) has been reported [2]:

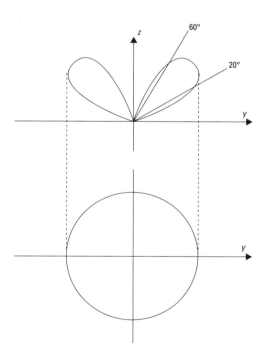

Figure 6.3 Receiving antenna radiation pattern for the mobile terminal.

$$A_F(\text{dB}) = -\left(3.44 + 0.0975 \cdot \text{EL}° - 0.02(\text{EL}°)^2\right)$$
$$\cdot \ln(p\%) + \left(-0.443 \cdot \text{EL}° + 34.76\right) \tag{6.6}$$

It may be used at frequencies near 1.5 GHz with elevation angles ranging from 20° to 60° and service outage probabilities between 1% and 20%.

Example 6.4

Determine the attenuation due to shadowing in a rural-suburban mobile environment in the L band if a service outage probability of 2% and a 50° elevation angle are intended.

Solution. From (6.6), obtain

$$A_F = -\left(3.44 + 0.0975 \cdot 50° - 0.02(50°)^2\right)\ln(2)$$
$$+ \left(-0.443 \cdot 50° + 34.76\right) = 10.3 \text{ dB}$$

The above attenuation can be compensated for by increasing the received power in the mobile receiver.

Because receiver motion induces rapid variations in signal level, temporal diversity techniques such as interleaving and channel coding must be used to combat short intermittent fading events.

The direct component has a possible Doppler frequency shift due to terminal motion. This frequency shift is given by

$$f_d = \pm\left(\frac{v}{c}\right)f_0 \cos(\text{EL}°) \tag{6.7}$$

where f_0 is the transmitted carrier frequency, v is the vehicle velocity, c is the speed of light, and EL° is the elevation angle of the line-of-sight in Figure 6.1.

Example 6.5

Determine the Doppler frequency shift for a mobile receiver receiving signals from a satellite DAB at 1.5 GHz. The elevation angle is 50°, and the receiver speed is 100 km/hr.

Solution. Using (6.6),

$$f_d = \pm \left(\frac{100 \cdot 10^3/3{,}600}{3 \cdot 10^8} \right) \cdot 1.5 \cdot 10^6 \cdot \cos 50^\circ = \pm 89.8 \text{ Hz}$$

The Doppler shift, plus any inherent local oscillator offset, represents the carrier bandwidth shift when received at the mobile terminal. Since the fading with a rapid Doppler frequency shift makes the problem of carrier recovery much more difficult, various techniques for coherent detection have been studied. Differential detection is also considered to be a simple and attractive method for recovering data from a fading signal, even though the static performance is inferior to that of coherent systems.

In the urban mobile environment, slower vehicle speed and blockage by buildings causes signal fades that are too long and too deep. The ATS-6 satellite was used to measure UHF and L-band excess path losses in various cities in the United States. As an example, in Denver, the satellite appeared in the southwest at a 32° elevation. A mobile running in a street perpendicular to the satellite direction experienced fades greater than 29 dB 10% of the time. This latter condition exceeds the mitigation capacity of any reasonable combination of power margin and channel coding.

Thus, in an urban mobile environment, the direct or line-of-sight component is practically absent because it is highly attenuated or blocked by objects such as buildings and bridges. Therefore, the multipath components are dominant.

Land-based signal boosters (Figure 6.4), also known as gap fillers, are needed to fill in the coverage gaps of the satellite. OFDM, spread spectrum, equalization, and other techniques are capable of compensating for channel effects and yielding significant performance improvements.

6.4 Transmission Techniques in Satellite DAB Systems

6.4.1 Sound Source Coding

DAB is based on a highly efficient form of source coding. Using the ISO/MPEG audio-coding techniques, compressed audio sources present the following characteristics:

- 16 Kbps for a monophonic source (AM quality);
- 32 Kbps for a monophonic source (FM mono quality);
- 64 Kbps for a stereophonic source (FM stereo quality);

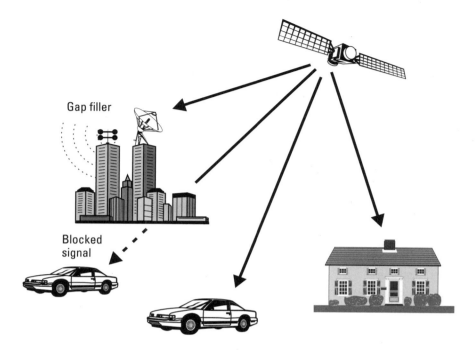

Figure 6.4 Terrestrial gap filler for DAB reception in an urban mobile environment.

- 128 Kbps for a stereophonic source (CD quality);
- The current audio standard ISO MPEG 2 layer 3 (MP3) or ISO MPEG 2 *advanced audio coding* (AAC) provides stereo CD quality at 96 Kbps, which is 10 times better than the CD standard.

6.4.2 Satellite Orbits

Often, portable and mobile receivers are not necessarily placed in the best position for good reception. Signals coming from a satellite are more likely to be obstructed when the satellite is at low elevation angles relative to the receiver. Low latitude coverage areas achieve satisfactory satellite visibility with a single geostationary satellite (e.g., WorldSpace) because of the relatively high elevation angle from the portable/mobile ground terminal to the satellite. For satellites that are low in the sky, a large margin is needed to compensate for the attenuation caused by trees, bridges, and buildings. This makes the geostationary orbit inappropriate for countries at higher latitudes. Possible solutions for coverage areas at higher latitudes include the use of several satellites in HEOs, providing high elevation angles over the coverage area (e.g., Sirius), or the use of at least two geostationary satellites at different orbital locations providing space diversity (e.g., XM). The Molnya or Tundra type of HEO provides a satellite that appears to be nearly stationary and overhead for several hours. By using a constellation of such satellites, it is possible to provide continuous coverage. The benefit of this is that for reception in cars, there is little blockage from obstacles like buildings or trees.

6.4.3 Modulation Scheme

In contrast to terrestrial transmission, modulation for satellite transmission does not need to be protected against multipath effects (except in an urban mobile environment where gap fillers are needed). Consequently, efficient modulation schemes such as TDM/QPSK can be used for satellite broadcasting. Several years ago, *coded orthogonal frequency division modulation* (COFDM) was specially developed for terrestrial transmission. It is an efficient type of modulation, adapted to multipath environments, that is used for emerging digital terrestrial broadcasting systems such as the ITU system A. However, TDM/QPSK modulation has a clear advantage over COFDM modulation for satellite applications. The comparison between ITU system A (COFDM plus Viterbi FEC coding) and TDM/QPSK transmission using concatenated Reed-Solomon and Viterbi FEC codes shows the following:

- The need of COFDM modulation to operate in linear mode (to avoid intermodulation effects) as it uses multicarrier signals, leading to RF power transmission below the maximum possible RF power and nonlinearity losses (about a 3-dB difference);

- The use of noncoherent demodulation for COFDM, as well as less efficient FEC (about a 4-dB difference);

- Losses inherent to COFDM modulation, that is, guard time and overhead allowance (about a 1-dB difference).

The results show an overall link margin difference of about 8 dB in favor of TDM/QPSK versus COFDM [3].

6.4.4 Diversity Techniques

The diversity technique, which is the most powerful method to combat fading, is widely used in various present-day microwave systems. From various diversity schemes, the space diversity is the most suitable one for satellite radio broadcasting, since without requiring any additional frequency spectrum it can keep the received signal at a higher level—even for deep fade. In satellite radio applications, space diversity is used at the transmitting side using various (typically two) satellites (Figure 6.5).

In diversity techniques, the uncorrelated faded signals received from the possible transmission path can be combined via the following three methods: maximal-ratio combining [Figure 6.6(a)], equal-gain combining [Figure 6.6(b)], and selection [Figure 6.6(c)].

Maximal-ratio combining gives the largest improvements in the performance of the three but results in the most complicated implementation. Equal-gain combining omits the weighting circuitry and still achieves comparable performance. Selection (or switching) is most suitable for mobile radio applications. The performance of these combining techniques is given by the following expressions, where it is assumed that Rayleigh fading affects each transmission path (urban mobile environment).

The diversity gain (G_d) can be defined as

$$G_d (\text{dB}) = A_F (\text{dB}) - A'_F (\text{dB}) \qquad (6.8)$$

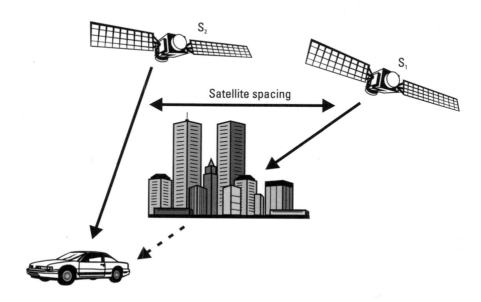

Figure 6.5 Space diversity used in satellite DAB systems.

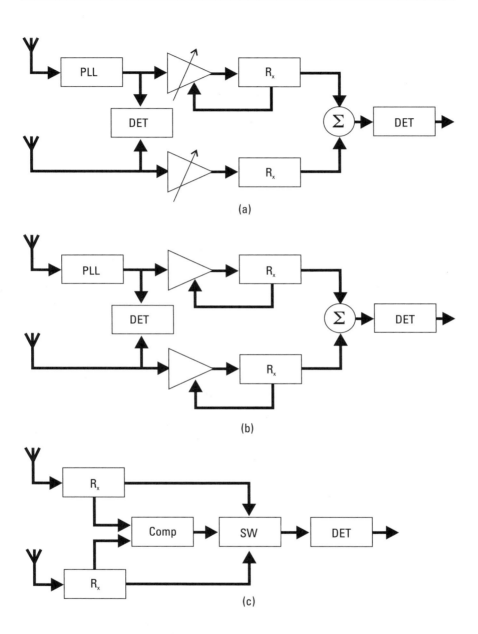

Figure 6.6 Typical combining methods: (a) maximal ratio combining, (b) equal-gain combining, and (c) selection (or switching).

where A_F' is the fade attenuation when two space diversity uncorrelated transmission paths are used and A_F is the fade attenuation without diversity. Both are calculated for the same outage performance.

It is important to point out the following aspects of diversity in satellite DAB systems:

- Transmitting satellites should have a minimum spacing in the geostationary orbit to guarantee that signal paths have no statistical correlation (uncorrelated fading paths).

- The receiver usually uses a single omnidirectional antenna. After reception, the receiver splits the incoming signal into two or more uncorrelated signal paths that have different delays.

6.4.5 Downlink Margin Equation for Satellite DAB Systems

The downlink margin equation for digital radio satellite systems is very similar to that for digital TV satellite systems, without any concern about rain conditions because of the characteristics of the operating frequency band. Using a nonregenerative transponder, it is possible to write

$$M_F = \text{EIRP} + G\,/\,T - L_b - \Sigma L - R_b$$
$$- \Delta N_U - \left(E_b\,/\,N_0\right)_0 + G_C + 228.6,\text{dB} \tag{6.9}$$

where the logarithmic units of each term have been omitted to easily write the equation. The left member in (6.9) is the fade margin, which, like its counterpart in satellite TV systems, is a power margin to guarantee the required outage performance. The fade margin M_F can be defined by the inequality

$$M_F > A_F, \text{dB} \tag{6.10}$$

where A_F is the fade attenuation that can be calculated using (6.6) for the rural-suburban mobile environment.

For satellite DAB systems that use space diversity techniques, the diversity gain G_d can be included in the following way:

$$M_F > A_F + G_d \tag{6.11}$$

6.5 Satellite DAB Projects

This section describes some of the main features of satellite DAB projects like Sirius, XM, and WorldSpace. As these DAB companies now have their space segments in place and have either finalized, or are close to finalizing, their satellite radio chipsets, each is engaged in the operational side of the business, performing such tasks as securing receiver manufacturing commitments and firmly establishing distribution and marketing strategies.

XM and Sirius will be broadcasting music from space via satellites. Both systems are going to deliver music exclusively to U.S. listeners. Sirius is using conventional satellites (with nonregenerative transponders) in HEOs. WorldSpace will broadcast to Africa, Asia, and Latin America using geostationary satellites with onboard processing. Table 6.1 compares the technical characteristics of the main satellite DAB systems today. Sections 6.5.1–6.5.3 will discuss these in more detail.

Table 6.1
Digital Audio Broadcast System Comparison

System Parameter	WorldSpace	Sirius	XM	Global Radio
Downlink band (MHz)	1,467–1,492	2,320–2,332.5	2,332.5–2,345	1,467.5–1,492
Band occupancy	2.5 MHz of above			
Tx HPA output (W)	300			
Uplink band (MHz)	7,025–7,075	7,025–7,075	7,025–7,075	?
Band occupancy	228 x FDMA- > 3 x TDM	?	?	?
OBP	Yes	No	No	?
Regions/ coverages	Three African spots @ 21° E => four streams Three Asian spots @ 105° E => six streams	CONUS	CONUS	Europe-Spots

Table 6.1 (continued)

System Parameter	WorldSpace	Sirius	XM	Global Radio
	Three Caribbean spots @? W => ? streams			
Satellites	Three GEOs, one per region	Three HEO 24-hour orbits	Two GEOs	Three HEO 12-hour orbits
S/C life (years)	15			?
S/C separated mass				?
S/C PL power (kW)	~4	~10	15	?
Terrestrial repeaters	0	Hundreds	> 1,000	?
Rx G/T (dB/K)	−13	?	?	?
Modulation coding	MPEG2/Layer 3	MUSICAM	PAC	
Transport coding	RS255/223 +? rate conv			
Transport	4 x QPSK	CDMA	6 x TDM/QPSK	S-DAB
Minimum stream rate	16 Kbps	Variable?	64 Kbps fixed	
Channels	3 x 96 x 16 Kbps		4 x 2 2.5-MHz carriers	S-DAB
User audio streams	~25/satellite	100	100	60–70
Radios	STARMAN			
LSI chips/ receiver	1?	> 8 (Initially)	2 (BBP & Src coder)	TBD
Chip technology (microns)	0.8			
Full service	2002?	2002?	May 15, 2001	
System cost	$700M?			

6.5.1 Sirius Satellite Radio

This system (formerly CD Radio) started broadcasting in 2001 to the continental United States. Sirius uses three SSL/1300 satellites in an inclined elliptical satellite constellation that provides elevation angles of 60° to 90° over the contiguous United States and ensures that each satellite spends about 16 hours a day over the continental United States, with at least one satellite over the country at all times. Sirius completed its three-satellite constellation on November 30, 2000. A fourth satellite is on the ground, ready to be launched if any of the three active satellites encounters problems.

A network of terrestrial transmitters that will broadcast the satellite signal in dense urban areas will also carry Sirius's broadcast signal. Currently, the terrestrial repeater network is 95% complete. Upon assessment of the signal strength of the three in-orbit satellites, Sirius decreased the total number of repeaters to 92 sites from the approximately 106 previously expected.

Sirius has three types of receivers available, depending on the requirements of the customer:

- Adaptive, or FM modulator, receiver with satellite module;
- Three-band integrated head unit receiver with satellite module;
- Three-band direct access integrated head unit receiver without the satellite module.

The Sirius receiver will include two parts, the antenna module and the receiver module. The antenna module will pick up signals from the satellite or ground repeaters, amplify the signal, and filter out any interference. The signal will then be passed on to the receiver module. Inside the receiver module will be a chip set that will convert the signals from 2.3 GHz to a lower IF.

The adaptive FM modulator version of the satellite receiver module allows a subscriber to retain his/her original car stereo and minimizes wiring in the host car's interior, making for a quicker installation. The satellite signal is received by a special roof-mounted satellite antenna and delivered to the satellite receiver module, which could be in the trunk of the vehicle. The satellite receiver module outputs via an FM modulator that utilizes an unused FM station on the existing FM receiver of the car to deliver the sound. All output (digital display) and input instructions are transmitted through a separate handheld device. This product gives the customer more flexibility and increases the size of the addressable market. However, it delivers sound quality limited to that of a clear analog FM signal. The three-band integrated head unit receiver, like the adaptive receiver, also requires a satellite module, but it has three bands: traditional AM and FM as well as satellite bands. Consequently, the sound is truly digital coming directly from the satellite feed, and the receiver has an attractive digital display. This product requires the original radio unit be specially manufactured as *satellite-ready*, meaning that it is compatible with the satellite module. The three-band direct access integrated receiver will only become available when the chipset has been shrunk sufficiently so that the chipset can be manufactured directly in the head unit itself. This product should retain all the benefits of the original three-band product and obviate the need for a satellite module, resulting in a potential cost savings to the consumer. These units are likely to be manufactured as *original equipment manufacturer* (OEM) equipment for the car companies and installed at the factory. In terms of availability, certain manufacturers, such as Kenwood, have already been selling tens of thousands Sirius satellite-ready receivers that are compatible with the satellite module. The satellite module was not expected to be available until the Agere chipset (Lucent) had been delivered and fully integrated into a receiver. The satellite module (including the satellite antenna and handheld display unit) will probably cost approximately $300–$400. The receivers are also going to be produced by Sanyo, Pioneer, Panasonic, Jensen, Clarion, and Alpine. Additionally, by 2004 Sirius intends to have developed, in conjunction with XM, an interoperable three-band direct access integrated head unit capable of receiving both Sirius and XM radio signals (200-channel availability).

Sirius plans to commence testing in conjunction with the completion of its validation tests for its chipset. Testing will be conducted in selective markets based on demographics, geography, and retail presence and will focus on the consumer experience. This process will incorporate all facets of the business, including programming, the repeater network, the acquisition experience, installation, and back-office processes such as customer activation and credit card processing. Currently, Sirius plans to charge subscribers $12.95 per month. A typical professional installation of the Sirius hardware in the vehicle will be $50–$100.

6.5.2 XM Satellite Radio

XM Satellite Radio (formerly American Mobile Radio) was broadcasting to the continental United States by November 2001. In just 6 months, XM has come from behind to lead Sirius Satellite Radio on many fronts. XM subscribers are expected to rise from 500,000 to 600,000, and comparable projections for Sirius have been lowered from 500,000 to 400,000. By June 2001, XM successfully launched its satellite constellation (two of two satellites), finalized its radio chipset, raised $200 million in the capital markets, lined up radio production, geared up retail aftermarket distribution, and secured factory-installed OEM distribution with GM.

The space segment of XM uses two Boeing HS 702 satellites, called "Rock," launched in May 2001, and "Roll," launched in June 2001, placed in geostationary orbit, one at 85° W longitude and the other at 115° W longitude, respectively. They are equipped with a payload supplied by Alcatel Space and are the most powerful communication satellites ordered to date (commercial operation has not been implemented as of this writing). Equipped with high-efficiency solar arrays and xenon ion thrusters, these 4.45-ton launch mass satellites will produce over 15.5 kW of power by the end of their 14-year working life. The payload comprises two nonregenerative transponders, each of which includes an antenna equipped with a shaped, single source, 5-m diameter reflector. Beam coverage is adapted to the shape of the North American continent, with a high EIRP, toward areas with low elevation angles as well as those with high population densities. Each transponder also includes a high-power amplification system composed of sixteen 216-W TWTAs operating in parallel at saturation. This antenna/HPA combination produces a peak EIRP of over 68 dBW. Figure 6.7 shows a panoramic view of the XM satellite radio system.

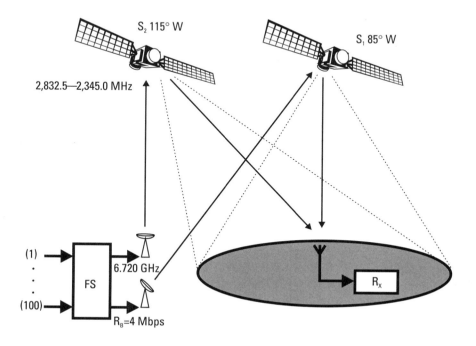

Figure 6.7 XM satellite radio system (FS: feeder station).

XM provides reception of audio and news, as well as text and other digital data on all types of receivers. The features of this new radio service include coast-to-coast coverage ensuring a national audience throughout the United States, high reception quality, and a very wide choice of programs (up to 100 programs). The song title, artist, and other topics will be displayed on the radio. XM will also complement its satellite system with a terrestrial repeater network to fill in coverage gaps caused by tall buildings that can block out the satellite signal. The radio antenna must have a direct view of the satellite to ensure that signals are received reliably. The repeaters, placed on rooftops and tower blocks, will receive and amplify the satellite audio signals and transmit them at much higher power levels. Each satellite radio receiver will use a small, car phone–sized antenna to receive the XM signal. Upon testing the signal strength of its two high-power geostationary satellites, XM decreased the total number of repeaters from 1,500 to approximately 1,300. It is worth noting that Sirius's repeater system, with only 92 sites, is significantly less complex.

One of the primary reasons for XM's solid execution has been the support of its strategic partners, including GM, DirecTV, and Clear Channel. The other significant difference is that Sirius has no commercials on its music channels, a feature that allows a lower price for XM's subscribers. General Motors (GM) has invested about $100 million in XM, and Honda has also signed an agreement to use XM radios in its cars. GM was scheduled to install XM satellite radio receivers in selected automobile models in 2002. For $9.95 per month, subscribers receive the XM signal. For that price, listeners will get up to 100 channels of music, talk, and news. Many of the channels have no commercials, and none of the channels air more than 7 minutes of advertising per hour. XM's content providers include *USA Today,* BBC, CNN/*Sports Illustrated,* and The Weather Channel. It will reinforce that lineup with its own music channels.

The XM satellite radio system uses the 2,332.5–2,345.0-MHz frequency band. It has been optimized for this S-band spectrum to ensure reliable performance in both urban and rural environments throughout North America. Flexible TDM/O-QPSK is used to broadcast music and voice. Each satellite transmits the same contents, so that a receiver can receive the contents from either of the satellite signals (maximal-ratio combining). Each receiver will contain a proprietary chipset (designed by ST Microelectronics) and a small, car phone—sized antenna to receive the signal. Figure 6.8 shows the receiver block diagram.

Like Sirius, XM has an adaptive receiver and integrated direct access three-band receivers. Adaptive products include Pioneer and Alpine's FM-

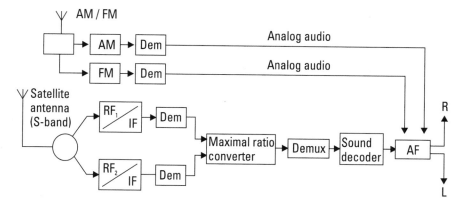

Figure 6.8 XM satellite radio receiver system.

modulated receivers, which can plug into the existing radio. One product that will differentiate XM is its Sony Plug-n-Play receiver. This receiver, which won awards at the Consumer Electronics Show in January 2001, is designed to allow the consumer to insert the satellite receiver into cradle receptacles located in both the vehicle and the home. With the satellite signal strength from its two satellites exceeding prior expectations, and its more extensive terrestrial repeater network, XM has greater faith in the potential of the home/portable market than Sirius. According to XM, the Pioneer adaptive receivers will retail for approximately $250–$300, with the comparable Alpine receiver expected to be priced slightly higher. The Sony Plug-n-Play is expected to retail for $300, including an antenna. Most three-band XM satellite receivers will retail for about $150–$200 over the retail price of a similar model, nonsatellite radio. This additional cost will purchase the satellite module and antenna. As with Sirius, XM subscribers also have to pay $50–$100 for a typical professional installation

6.5.3 WorldSpace

WorldSpace is a North American company that has implemented a satellite digital radio broadcasting system for underdeveloped areas around the world. In 1995, Alcatel Espace signed a contract with WorldSpace for the turnkey delivery of the entire system. The space segment was completely manufactured by Alcatel Espace at its facility in Toulouse, France. The first satellite (AfriStar) was placed in orbit in October 1998 and is broadcasting more than 50 radio programs throughout the African continent and to parts of Europe.

AsiaStar was launched in March 2000. WorldSpace will be able to broadcast to the majority of the world's population when its AmeriStar satellite is launched.

Initially, the United States will not be part of WorldSpace's coverage area. However, the company has invested in XM and has an agreement with the company to share any technological developments. WorldSpace does plan to reach parts of the world that most radio stations cannot. There are millions of people living in WorldSpace's projected listening area that cannot pick up a signal from a conventional radio station. WorldSpace has a potential audience of about 4.6 billion listeners spanning five continents.

WorldSpace is the first satellite digital radio broadcasting system to provide reception on portable radio sets. It offers coverage over developing and underdeveloped countries using geostationary satellites and L-band frequency transmission. The service is mainly targeted at regions where radio broadcasting is still practically nonexistent and where low-cost, portable radios are urgently required. The service also supports data transmission (e.g., maps giving weather forecasts or road traffic information). A number of dedicated applications are under development.

The WorldSpace system consists of three geostationary satellites:

- AfriStar, at 21° E orbital location, provides DAB to Africa, the Middle East, and part of Europe and was launched in October 1998.

- AsiaStar, at 105° E orbital position, provides DAB to Southeast Asia and the Pacific Rim and was launched in March 2000.

- AmeriStar, at 95° W orbital position, provides DAB to Central and South America and is in preparations for launch at the time of this writing.

Each satellite belongs to a standard large class with the following primary characteristics:

- Launch mass: 2,750 kg;
- Satellite bus: EuroStar 2000, by Astrium (formerly Matra-Marconi Space);
- Power: Solar cells (5,550W) and backup batteries;
- Operation: sun and eclipse (24 hours);
- Station-keeping stability: ± 0.1° in N/S and E/W directions;
- EIRP: 49.8 dBW;

- $(G/T)_S$: −12.6 dB/K (global beam);

- Uplink coverage area: global beam;

- Downlink coverage area: three spot beams per satellite;

- Satellite dimensions: $(2.6 \times 1.8 \times 2.3)$m.

- Lifetime: 15 years.

The system configuration for each satellite is shown in Figure 6.9. The satellites are operated by a ground control segment and managed, according to the traffic requirements, by a mission segment for the operating lifetime of the system.

The WorldSpace system uses the frequency band allocated at WARC'92 for BSS DAB—that is, 1,467–1,492 MHz, in accordance with ITU Resolutions 33 and 528. Broadcasters use VSAT-type feeder uplinks transmitting in the X band from 7,025 to 7,075 MHz.

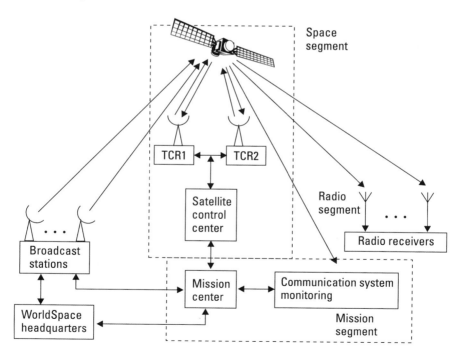

Figure 6.9 WorldSpace system configuration.

The communication payload enables all potential broadcasters to have direct access to the satellite using FDMA. Uplink signals issued by the broadcasters are transmitted via individual FDMA channels from Earth stations located anywhere within the view of the satellite with elevation angles higher than 10°. FDMA offers the highest possible flexibility between multiple independent broadcasting stations, allowing each broadcaster to send a signal from its own facilities directly to one of the WorldSpace satellites.

Broadcast stations transmit their digital audio programming directly to the satellite, which broadcasts the signal to the radio receivers. The mission segment manages access to the satellite by the broadcasters.

The WorldSpace coverage area is shown in Figure 6.10. Three downlink spot beams cover each region. Each WorldSpace satellite has a wide coverage uplink beam (global beam) that contains the three-downlink spot beams that operate in the X band and use a different onboard receiving antenna. This uplink global beam is not shown in Figure 6.10.

Depending on the user's location within the coverage area, the use of geostationary satellites offers high elevation angles for radio program reception (generally 50° or higher). Such high elevation angles limit shadowing effects and avoid multipath that might affect the signal of the satellite broadcasts.

Each satellite can transmit a total capacity of 1,536 Kbps per TDM, which might be any mix of the audio services (16-, 32-, 64-, or 128-Kbps). This corresponds to a capacity per beam of one, or a combination, of the following signal types:

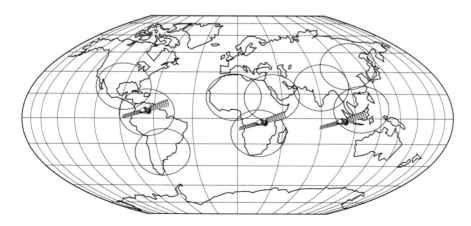

Figure 6.10 WorldSpace coverage areas.

- 96 mono (AM quality) audio channels;
- 48 mono (FM quality) audio channels;
- 24 stereo (FM quality) audio channels;
- 12 CD quality stereo audio channels.

At the studio, the broadcaster multiplexes the radio programs into a broadcast channel with a bit rate of 16–128 Kbps, depending on the quality and number of programs. Prior to transmission to the satellite, the broadcasting station time demultiplexes the broadcast channel into prime-rate 16-Kbps channels that are transmitted using FDMA. The overall uplink capacity is 288 prime-rate channels.

Each downlink transmits 96 prime-rate channels of 16-Kbps, TDM data in a carrier with a bandwidth of approximately 2.5 MHz. Each of the three beams uses a different carrier frequency set and is flexible within the frequency band. The satellite payload transmits each TDM at saturation, giving the highest possible power efficiency.

Conversion between FDMA and TDM is carried out onboard the satellite at the baseband level. Baseband processing provides a high level of channel control, such as channel routing between the uplink and the downlink, thus allowing FDMA frequency allocation on the uplink to be flexible. See Figure 6.11 for a depiction of the operational concepts.

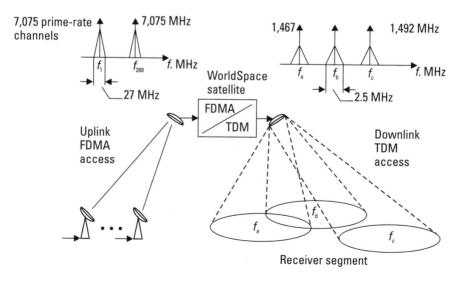

Figure 6.11 Communication operations.

The modulation scheme used in the WorldSpace system is QPSK with concatenated channel coding. There is no plan to use terrestrial gap fillers in the WorldSpace system. Signal quality and availability depend on the satellite only.

The WorldSpace satellite payload uses baseband processing that has already been proven on advanced technology satellites but that is only now being used in commercial satellite projects. Baseband processing improves system performance in terms of the uplink and downlink budgets, the management of broadcasting stations, and the control of the downlink signals. In the case of WorldSpace, it makes the broadcasters' uplink RF hardware much smaller and less expensive. This is consistent with having numerous distributed uplinkers and contrasts with the centralized uplinks envisioned for use with Sirius and XM. Figure 6.12 is a block diagram of the WorldSpace communication payload baseband processing.

The baseband processor demodulates all the received uplink FDMA prime-rate channels. For each downlink TDM signal, 96 uplink prime-rate channels are selected from among all the uplink signals and then time-multiplexed in the downlink. The selected prime-rate channels correspond to several broadcast channels that will be received by the radio receivers.

The baseband processor allows flexible channel management:

- Any uplink channel can be broadcast at the L band simultaneously to one, two, or three beams.

- The baseband processor monitors each uplink channel.

- Selection of uplink channels within each downlink TDM is controlled continuously.

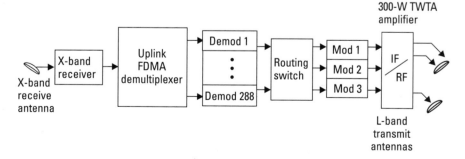

Figure 6.12 WorldSpace satellite communication payload.

Feeder link stations use VSAT terminals to transmit between one and several prime-rate (16-Kbps) uplink channels using small antennas (2–3m diameter) and low-power amplifiers (10–100W of RF power). Ground equipment for digital MPEG audio coding can be located either in the broadcasting station or in the broadcaster's studios. Figure 6.13 is a block diagram of a feeder link station.

Hub broadcast stations, which can transmit a large number of channels, are available. Links between broadcasters and hub stations use standard PSTN communication links.

WorldSpace's radio receivers are based on mass-produced ASIC chips that can be easily integrated into multiband radios to ensure they meet the target cost. They can operate with solar power or batteries. WorldSpace has contracted four consumer electronics companies to produce small portable receivers, including JVC, Matsushita (Panasonic), Hitachi, and Sanyo. Receiver costs will range between $250 and $550. Each receiver will be capable of receiving data at a rate of 128 Kbps. The receivers will use the proprietary StarMan chipset manufactured by ST Microelectronics to receive digital signals from the satellites. Figure 6.14 shows the block diagram of the radio receiver.

The radio receives the L-band signal, demodulates and extracts the useful audio signal from the TDM stream, and decodes it into its original user format.

The radio is equipped with a simple moderately directional array antenna for outdoor reception and most indoor reception. In certain

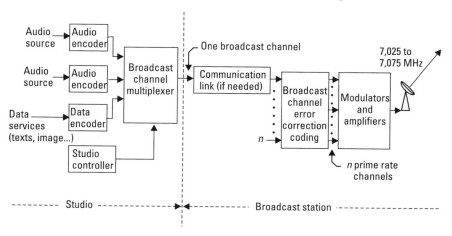

Figure 6.13 Feeder link station.

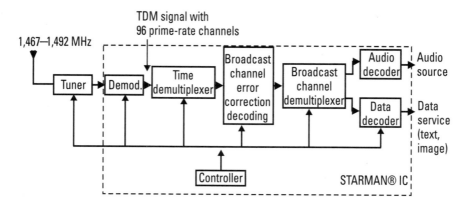

Figure 6.14 WorldSpace radio receiver.

environments (concrete buildings or buildings with metal structures), an external antenna may be needed to provide adequate indoor reception. Because the system uses digital techniques, the radio can receive ancillary multimedia services, including slow-motion video, paging, mail, and fax. These can be presented on a flat-panel display or forwarded over a serial data interface to a PC. This data is multiplexed within the digital audio signal channels. The system fully supports encryption by the broadcasters.

WorldSpace subscription details will be determined by local distributors, but there will be multiple service options, ranging from a low-cost basic service to a high-end specialty service. WorldSpace said that the cost for its service would likely be a fraction of dial-up Internet costs. Discover, CNN, BBC, Bloomberg News, and MTV Asia will deliver programming for WorldSpace, among others.

The main radio receiver parameters are listed as follows:

- Operating frequency band: 1,467–1,492 MHz;
- Polarization: circular;
- Antenna radiation pattern: toroidal;
- Receiver noise figure: 1.4 dB;
- Demodulator implementation losses: 1.8 dB;
- Antenna gain: 9–12 dBi;
- Receiver (G/T): −13.4 to −16 dB/K;
- Antenna noise temperature: 90–100K.

The WorldSpace broadcasting system will deliver digital signals with a BER of 10^{-4} or better, providing the various service qualities defined earlier.

Each downlink beam will have a bandwidth of about 6°, offering a link margin of 10 dB, assuming a −13.4-dB/K gain-to-temperature ratio radio receiver. This margin will help to combat signal loss caused by obstacles in the path between the satellite and the receiver, providing full-quality reception within the coverage area. The margin equation for each downlink beam can be written as

$$M_F = \text{EIRP} + G / T - L_b - \Sigma L$$
$$-R_b - \left(E_b / N_0\right)_0 + G_c + 228.6, \text{dB} \qquad (6.12)$$

where the term ΔN_U has been neglected because of the regenerative characteristics of the WorldSpace satellite communication payload.

Extended coverage will be possible in unobstructed open areas using the standard radio receiver. In areas where reception conditions are less favorable, the radio receivers can be connected to large-gain antennas or to an antenna located in an unobstructed position. For example, reception in a large building might require a common roof-mounted antenna for the entire building or individual antennas mounted close to windows.

Example 6.6

Calculate the satellite EIRP for AmeriStar satellite (95° W) if the mobile receiver is located near the coordinates 10° N, 70° W. Assume the following parameters for the mobile receiver:

- G/T = 13.4 dB/K;
- BER of 10^{-5} for 1% service outage (yearly average);
- Coherent QPSK;
- Coding gain: 7 dB;
- Detector implementation loss: 1.8 dB;
- Reception environment: rural-suburban.

Solution. The free-space loss is

$$L_b = 185 + 20 \log(1.5) + 10 \log\left(1 - 0.296 \cos\left(-25°\right) \cdot \cos\left(10°\right)\right) = 187.2 \, \text{dB}$$

The information bit rate in decibels per bits per second is

$$R_b = 10 \log(96.16 \cdot 10^3) = 62 \text{ dB.bps}$$

For a 10^{-5} BER, the corresponding $(E_b/N_0)_0$ value is 9.6 dB. The elevation angle is

$$EL° = \arctan$$

$$\left\lfloor \left(\cos(-25°).\cos(10°) - 0.1513\right) / \left(\sqrt{1 - \cos^2(-25°).\cos^2(10°)}\right) \right\rfloor = 59°$$

The fade attenuation A_F is

$$A_F = -0.443 \cdot 59° + 34.76 = 8.6 \text{ dB}$$

Using (6.10), then, it is possible to select an M_F value of 10 dB. Using (6.12), the EIRP value is

$$EIRP = 10 + 13.4 + 187.2 + 1.8 + 62 - 7 + 9.6 - 228.6 = 48.4 \text{ dBW}$$

and it represents a good EIRP estimation (the nominal value is 49.8 dBW).

Example 6.7

Assume now that the WorldSpace system uses a single downlink beam with a full capacity of 288 prime-rate channels instead of the 96 prime-rate channels per downlink beam. Using the same link parameters in Example 6.6, determine the new EIRP requirements for the transmitting satellite.

Solution. The new information bit rate is

$$R_b = 10 \log(288.16 \cdot 10^3) = 66.6 \text{ dB.bps}$$

Then,

$$EIRP = 10 + 13.4 + 187.2 + 1.8 + 66.6 - 7 + 9.6 - 228.6 = 53 \text{ dBW}$$

and the increment in satellite power is

$$53 - 48.4 = 4.6\,\text{dB}$$

and is a logical result.

References

[1]   Bell, D., et al., "Overview of Techniques for Mitigation of Fading and Shadowing in the Direct-Broadcast Satellite Radio Environment," *International Mobile Satellite Conference (IMSC'95)*, Ottawa, Canada, 1995, pp. 423–432.

[2]   Vogel, W., and J. Goldhirsh, "Mobile Satellite System Propagation Measurement at L-Band Using MARECS-B2," *IEEE Trans. on Antennas and Propagation*, Vol. 38, No. 2, February 1990, pp. 259–264.

[3]   Courseille, O., and P. Fournié, "WorldSpace: The World's First DAB Satellite Service," *Alcatel Telecommunication Review*, Second Quarter 1997, pp. 102–107.

7

Conclusions

Satellite communications have been evolving from a global business pushed forward by technology (bandwidth providers) to a mainly commercial one driven by user's needs. This new business scenario is now confirmed by a convergence of service providers and technology development, giving a higher level of competition. In this environment, satellite systems using geostationary platforms for broadcasting applications like satellite TV, satellite radio, and satellite Internet are reaching further down to the end user in this increase in demand for services. Digital broadcasting is an evolving technology that will provide an enhanced variety and choice of programs and a robust delivery mechanism. Today, a huge market, strong growth, a continuously evolving technology, and the use of worldwide standards for transmission and signal compression characterize digital satellite-broadcasting systems.

Significant improvements have been made in satellite technology. Satellites continue to become larger and more massive as launch capabilities increase and more efficient technologies are introduced. Geostationary broadcast satellites are not deployed uniformly around the orbit but are clustered over regions where services are more in demand. Shaped-beam technology is employed to cover service areas efficiently and to offer geographical isolation capability that is not feasible with simple beams. Satellite manufacturers are motivated to shorten the time needed to construct the satellite both to allow operators to implement their business plans earlier and to capture a reduced "marching army" cost (much of the cost of building a satellite is personnel). Using production line techniques and other best practices, manufacturers have successfully reduced the satellite construction period to less than 2 years in many instances. In addition, the need for greater capacity and service flexibility has supported the introduction of digital repeaters with OBP. Digital satellite TV systems (Skyplex) and digital radio satellite systems (WorldSpace) are among the first ones to employ digital signal processing in space.

Standards have played a vital role in digital television and radio systems. MPEG and other standards for signal coding have permitted the compression of audiovisual signals producing a lower bit rate signal. The economics of introducing signal compression methods has been a tradeoff between the cost of the processing electronics, the reduction in transmission rate (which reduces the cost of providing the service), and the user acceptance of the quality in the restored signal. High-speed semiconductor memories and microprocessors, now available, achieve a greater compression for the same expense.

The DVB standards have permitted the optimization of the satellite transmission system for TV programs in terms of power savings, service flexibility, and robustness against noise and interference. Proprietary systems such as DirecTV have done well initially but it seems likely that international standards (like DVB) will take over, sooner or later, in a global market. Use of a common TS multiplex (MPEG-2 TS) for satellite, cable, and terrestrial VHF/UHF networks has been fundamental for the success of the digital broadcast services in reaching the necessary commonality between different transmission media.

The digital satellite platform also provides an important opportunity to offer a variety of information services. The Internet is the most promising business in the field of satellite communications and broadcasting. The same receiver that provides more than 200 programs of compressed full-motion video is also an effective way to distribute Internet content. It is expected that Internet service providers will use satellite systems to complement the terrestrial infrastructure. They will also aim at emerging applications such as telemedicine and distance learning. These applications are going to take advantage of the point-to-area capability of satellites, serving multiple users simultaneously.

Reducing the cost of the receiving terminal is vital in broadcast applications. To be a success, any new digital broadcasting system has to have significant market penetration from the beginning, and a parameter of greatest importance is the cost of the receiver. As well as attaining the price target, the receiver must be as simple as possible to operate, allow access to a variety of services, deliver reasonable picture and sound quality, include a CA subsystem, and interface with ease to the consumer's existing television equipment. Fortunately, receiver cost reduction is much more an issue of manufacturing run size than technology improvement. The emphasis now is on using the smallest possible receiving antenna to reduce cost and simplify installation. In the 1990s services like DirecTV and Dish introduced low-cost digital receivers with 45-cm dishes, generating a very large market in the satellite industry. There is now a high degree of circuit integration (in the form of a chipset) to pack that amount of technology in a receiver. Video- and audio-decoding VLSI is now offered along with demultiplexing and FEC decoding on either the same chip or in a chip set. Suitable LNBs and tuner front ends are also available. In the near future software will play a predominant role in the receiver configuration. Operating code and receiver settings will probably be received via the satellite link. The next step is to add technology features (such as integrated hard drives) to facilitate the convergence between TV and computing. The broadband downlink can then deliver an arrangement of services to a superior receiver, and a lot of content will already be there when the user requests it.

There is a very strong interest in the new satellite systems dedicated to providing DAB; systems such as WorldSpace and XM Radio are going to provide DAB services in different parts of the world. Inexpensive mobile receivers will become available and this will prompt the takeoff of the service with global broadcasters. DAB systems are affected by several constraints like the limited spectrum available in the L and S bands, so it is critical to achieve compact antennas and a space segment cost reduction.

In the coming years, satellite systems will play a major role in facilitating the development of the information superhighway; they will do this through such avenues as advanced DTH broadcasting, global Internet access, and digital audio services. An adequate information infrastructure is undoubtedly an essential element for improving our quality of life, and satellite broadcasting is helping to sustain the development of all humankind.

Appendix A:
Elevation and Azimuth Angle Formulas

A.1 Elevation Angle

Let us consider triangle OTS in Figure A.1, which is reproduced as a planar projection in Figure A.2.

In triangle OTS, one can write the following trigonometric relation:

$$\tan(90°+\text{EL}°) = \frac{(R+H)\sin\beta}{R-(R+H)\cos\beta} \qquad (A.1)$$

Since

$$\tan(90°+\text{EL}°) = -\cot\text{EL}° \qquad (A.2)$$

then

$$\tan\text{EL}° = \frac{(R+H)\cdot\cos\beta - R}{(R+H)\cdot\sin\beta} \qquad (A.3)$$

Using the law of cosine in triangle OTS,

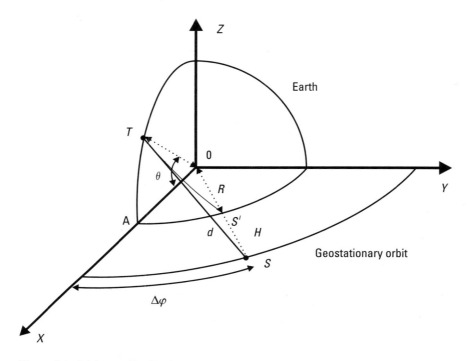

Figure A.1 Basic satellite-Earth station geometry.

$$d = \sqrt{(R+H)^2 + R^2 - 2R(R+H)\cos\beta} \qquad \text{(A.4)}$$

Comparing (A.4) with (3.10), it is possible to write

$$\cos\beta = \cos\Delta\varphi \cdot \cos\theta \qquad \text{(A.5)}$$

and $\Delta\varphi$ is defined as

$$\Delta\varphi = \varphi_S - \varphi_T \qquad \text{(A.6)}$$

where φ_S the satellite's orbital position and φ_T is the ground terminal longitude. θ is the ground terminal (T) latitude. North latitudes and east longitudes are going to be considered positive-signed; south latitudes and west longitudes are negative-signed.

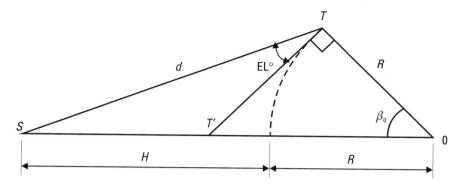

Figure A.2 Auxiliary projection for calculating elevation angle EL°.

Remembering that

$$\sin \beta = \sqrt{1 - \cos^2 \beta} = \sqrt{1 - \cos^2 \Delta\varphi \cdot \cos^2 \theta} \qquad (A.7)$$

then, substituting (A.5), (A.6), and the numerical values of R and H into (A.3), finally obtain (3.13):

$$EL° = \arctan\left(\frac{\cos \Delta\varphi \cos \theta - 0.1513}{\sqrt{1 - \cos^2 \Delta\varphi \cos^2 \theta}}\right) \qquad (3.13)$$

where

$$0 \le EL° \le 90° \qquad (A.8)$$

A.2 Azimuth Angle

Figure A.3 represents the spherical right triangle TAS'.

In this kind of triangle, each side is measured by its central angle. Thus, the length of arc AS' is measured by the angle $\Delta\varphi$, and the length of arc AT', by the angle θ. Using special rules, known as Napier's rules [1, 2], it is possible to write

$$\tan \Delta\varphi = \sin \theta \cdot \tan \gamma \qquad (A.9)$$

and then

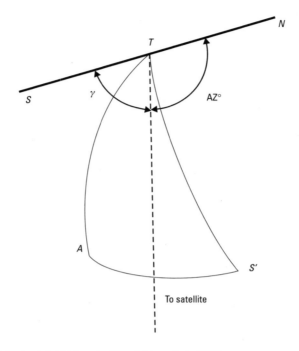

Figure A.3 Spherical right triangle. The right angle is TAS'.

$$\gamma = \arctan\left[\frac{\tan \Delta\varphi}{\sin \theta}\right] \tag{A.10}$$

The azimuth angle AZ° is

$$AZ° = 180° - \gamma \tag{A.11}$$

and finally obtain (3.14):

$$AZ° = 180° - \arctan\left(\frac{\tan \Delta\varphi}{\sin \theta}\right) \tag{3.14}$$

It should be observed that AZ° is measured (in the northern hemisphere) from the geographic north and rotating from the left of an ideal observer looking at the geographic south. When the ground terminal is in the southern hemisphere, the reference must be in the geographic south, and the other things remain the same.

References

[1]   Bronshtein, I., and K. Semendiaev, *Manual of Mathematics,* Moscow, Russia: Mir Publishing, 1971, pp. 220–222.

[2]   Pritchard, W. L., H. G. Suyderhoud, and R. A. Nelson, *Satellite Communication Systems Engineering,* 2nd ed., Englewood Cliffs, NJ: Prentice Hall, 1993, pp. 98–100.

Appendix B:
Calculation of Rain Attenuation Using the ITU-R's Method (Rec. ITU-R PN.837-1, 1992–1994; ITU-R PN.837 and PN.839-2, 94-97-99)

The ITU-R's method for calculating rain attenuation is good enough for general planning purposes. The step-by-step procedure is described as follows:

- Locate the receiving ground terminal on the map (Figures B.1, B.2, or B.3) and determine the rain climate region (A, B, C, D, E, F, G, H, J, K, L, M, N, P, or Q).

- Obtain the rain rate $R_{0.01}$, in millimeters per hour in Table B.1 and for 0.01% service outage (yearly average).

- Estimate the height of the 0°C isotherm during rainy conditions, h_{FR}, using

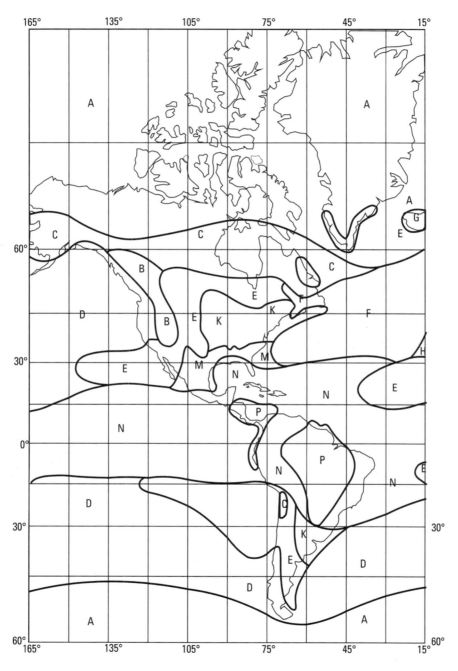

Figure B.1 Rain climate region: America.

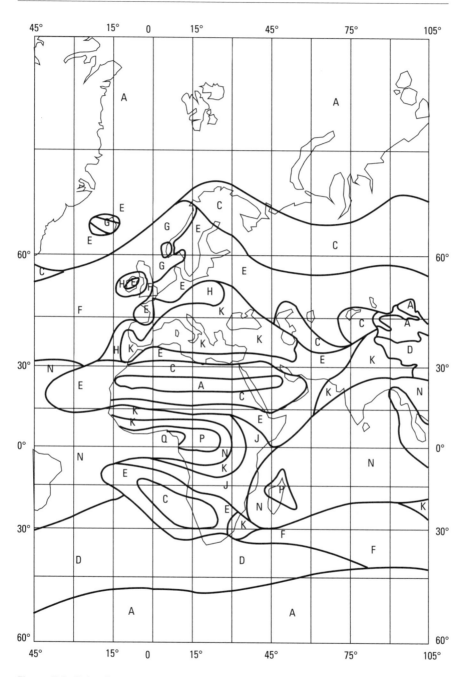

Figure B.2 Rain climate region: Europe-Africa.

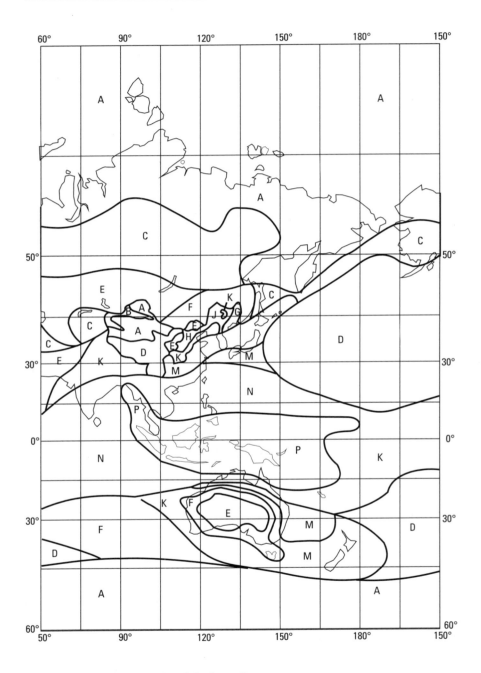

Figure B.3 Rain climate region: Asia-Australia.

Table B.1
Rain Intensity (Reference to Figures B.1–B.3)

Percent-age of Time (%)	A	B	C	D	E	F	G	H	J	K	L	M	N	P	Q
1.0	<0.1	0.5	0.7	2.1	0.6	1.7	3	2	8	1.5	2	4	5	12	24
0.3	0.8	2	2.8	4.5	2.4	4.5	7	4	13	4.2	7	11	15	34	49
0.1	2	3	5	8	6	8	12	10	20	12	15	22	35	65	72
0.03	5	6	9	13	12	15	20	18	28	23	33	40	65	105	96
0.01	8	12	15	19	22	28	30	32	35	42	60	63	95	145	115
0.003	14	21	26	29	41	54	45	55	45	70	105	95	140	200	142
0.001	22	32	42	42	70	78	65	83	55	100	150	120	180	250	170

$$b_{FR} = \begin{cases} 5 - 0.075(\theta - 23) & \text{for} & \theta > 23 \text{ NH} \\ 5 & \text{for} & 0 \leq \theta \leq 23 \text{ NH} \\ 5 & \text{for} & 0 \geq \theta \geq -21 \text{ SH} \\ 5 + 0.1(\theta + 21) & \text{for} & -71 \leq \theta \leq -21 \text{ SH} \\ 0 & \text{for} & \theta < -71 \text{ SH} \end{cases}$$

where NH and SH mean northern and southern hemispheres, respectively; b_{FR} is in kilometers, and θ is the latitude in degrees.

The slant path L_S in kilometers is

$$L_S = \frac{b_{FR} - b_S}{\sin \text{EL}°}$$

where EL° is the elevation angle and b_S is the height of the ground terminal above sea level, in kilometers.

The reduction factor $r_{0.01}$ due to the inhomogenity of rain, is given by

$$r_{0.01} = \frac{1}{1 + \dfrac{L_g}{L_O}}$$

where

$$L_g = L_S \cdot \cos \text{EL}°$$

and

$$L_O = 35 \cdot \exp(-0.015 \cdot R_{0.01})$$

The specific rain attenuation $\gamma_{0.01}$ in decibels per kilometer is

$$\gamma_{0.01} = k \cdot (R_{0.01})^{\alpha}$$

where coefficients k and α are given by

$$k = \left[k_H + k_V + (k_H - k_V) \cdot \cos^2 \text{EL}° \cdot \cos 2\tau \right] / 2$$

$$\alpha = \left[k_H \alpha_H + k_V \alpha_V + (k_H \alpha_H - k_V \alpha_V) \cdot \cos^2 \text{EL}° \cdot \cos 2\tau \right] / 2k$$

Table B.2
Regression Coefficients for Estimating Specific Attenuation (dB/km)

Frequency (GHz)	k_H	k_V	α_H	α_V
4	0.000650	0.000591	1.121	1.075
6	0.00175	0.00155	1.308	1.265
7	0.00301	0.00265	1.332	1.312
8	0.00454	0.00395	1.327	1.310
10	0.0101	0.00887	1.276	1.264
12	0.0188	0.0168	1.217	1.200
15	0.0367	0.0335	1.154	1.128
20	0.0751	0.0691	1.099	1.065
25	0.124	0.113	1.061	1.030
30	0.187	0.167	1.021	1.000
40	0.350	0.310	0.939	0.929

where τ is the polarization tilt angle relative to horizontal ($\tau = 0°$ for horizontal polarization; $\tau = 90°$ for vertical polarization; and $\tau = 45°$ for circular polarization).

The frequency-dependent coefficients k_H, k_V, α_H, and α_V are given in Table B.2. Values of k and α at frequencies other than those in Table B.2 can be obtained by interpolation using a logarithmic scale for frequency, a logarithmic scale for k, and a linear scale for α; that is,

$$\log k = \left(\log k_2 - \log k_1\right) \frac{\log f - \log f_1}{\log f_2 - \log f_1} + \log k_1$$

$$\alpha = \left(\alpha_2 - \alpha_1\right) \frac{\log f - \log f_1}{\log f_2 - \log f_1} + \alpha_1$$

These values have been tested and found reliable for frequencies up to about 40 GHz.

The effective length L_e in kilometers, in the rainy path is

$$L_e = L_S \cdot r_{0.01}$$

Rain attenuation $A_{0.01}$ in decibels is

$$A_{0.01} = \gamma_{0.01} \cdot L_e$$

For p% of service outage, then

$$\frac{A_{p\%}}{A_{0.01}} = 0.12 \cdot p(\%)^{-[0.546 + 0.043 \cdot \log(p\%)]}$$

Useful Formulas

For region 2 and $f = 12.5$ (the center of the downlink frequency band):

- Polarization: H/V

$$k = 0.020 \pm 1.08 \cdot 10^{-3} \cdot \cos^2(EL°)$$

$$\alpha = \lfloor 0.024 \pm 1.48 \cdot 10^{-3} \cdot \cos^2(EL°) \rfloor / k$$

- Polarization: Circular

$$k = 0.02; \; \alpha = 1.12$$

The relation between the worst month's probability outage (p_W) and the average yearly outage (p), both expressed in percentages, is as follows:

$$p = 0.3 \cdot \left(p_W \right)^{1.15}$$

Appendix C:
Coding Gain

A system designer is interested in quantifying the improvement in using channel coding in terms of allowable reduction of the output BER for the same C/N or the reduction of the C/N for the same BER. Usually, the term *coding gain* is introduced to denote the improvement in the performance of the system and is defined as follows:

$$G_C = (E_b / N_0)_U - (E_b / N_0)_C, \text{dB}$$

where $(E_b / N_0)_U$ and $(E_b / N_0)_C$ are the energy/bit figures relative to the noise power per hertz of the uncoded and coded frame, respectively, to give the same BER at a receiver. The assumption is that the bit rate of information transmission remains the same in both cases. This implies that the bit rate of the coded signal R_C is higher than for the uncoded case R_b. The relation between the energy of the coded bit E_c and the energy of the uncoded bit E_b is

$$E_c = r \cdot E_b$$

where r is the FEC code rate.

Example

Let a QPSK signal have $E_b/N_0 = 9.6$ dB. Determine the following:

1. The BER for the uncoded case;
2. The new BER if a BCH code is inserted with the following parameters:
   > Coded-word length (n): 63-bit;
   > Redundancy (r): 12-bit;
   > Error-correction capacity (t): 2-bit.
3. The increment in the bandwidth, in percentage, for the coded case relative to uncoded case;
4. The coding gain for the same BER as (1).

Solution.

1. For QPSK modulation, one can use (4.14) evaluated via (4.16):

$$P_b = Q\left(\sqrt{\frac{2E_b}{N_0}}\right) = Q\left(\sqrt{2.10^{0.96}}\right) = Q(4.27) = 10^{-5}$$

2. For the coded case with a BCH code with constant power, the E/N weakens:

$$\frac{E_c}{N_0} = r \cdot \frac{E_b}{N_0} = \frac{51}{63} \cdot 10^{0.96} = 7.38$$

The BER p_c for the coded case can be calculated as

$$p_c = Q\left(\sqrt{2 \cdot 7.38}\right) = 6.46 \cdot 10^{-5}$$

An approximated formula for calculating the output-decoded bit-error probability for the coded case is [1, 2]

$$P_b = \frac{1}{n} \sum_{j=t+1}^{n} j\binom{n}{j} p_c^j \cdot (1 - p_c)^{n-j}$$

where t is the largest number of channel bits that the code can correct within each block of n bits. Substituting numerical values, obtain

$$P_b = \frac{1}{63} \sum_{j=3}^{63} j \binom{63}{j} \left(6.46 \cdot 10^{-5}\right)^j \cdot \left(1 - 6.46 \cdot 10^{-5}\right)^{63-j} \approx 5.1 \cdot 10^{-10}$$

3. The bandwidth increment is

$$\Delta B\% = \left(r^{-1} - 1\right) \cdot 100 = \left(\frac{63}{51} - 1\right) = 23.5\%$$

Using the same formula as in step 2 and using an approximation, it is possible to write

$$10^{-5} = \frac{3}{63} \binom{63}{3} \cdot p_c^3$$

Then, p_c is $1.742.10^{-3}$. Using the formula for the BER in a QPSK system,

$$1.742 \cdot 10^{-3} = Q\left(\sqrt{\frac{2E_c}{N_0}}\right)$$

and obtain $E_c/N_0 = 6.387$ dB. Now, the new value of E_b/N_0 can be calculated as

$$\left(E_b \Big/ N_0\right) = \frac{\left(E_c \Big/ N_0\right)}{r} = \frac{10^{6.387/10}}{51/63} = 5376$$

and using decibels, $E_b/N_0 = 7.30$ dB. The coding gain G_C is

$$G_C = 9.6 - 7.3 = 2.3 \text{ dB}$$

References

[1] Sklar, B., "Defining, Designing, and Evaluating Digital Communication Systems," *IEEE Communications Magazine,* November 1993, p. 98.

[2] Odenwalder, J. P., *Error Control Coding Handbook,* San Diego, CA: Linkabit Corporation, 1976.

Appendix D:
Feed Loss Contribution to Antenna Noise Temperature

The feed loss L_f (in decibels) increases the antenna noise temperature when this parameter is measured at the feed's output. Let us consider the thermal noise model depicted in Figure 3.8, where now T_{sky}, T_{rain}, and A_R are substituted by T_A (antenna noise temperature without considering feed loss), T_f (feed noise temperature), and L_f, respectively. Following the same reasoning as that detailed in Section 3.8.4.2, the antenna noise temperature at the output of the parabola's feed is

$$T_{A/f} = 10^{-1.0 \cdot L_f} \cdot T_A + 290\left(1 - 10^{-0.1 \cdot L_f}\right) \text{K}$$

Example

Determine the antenna noise temperature at the feed output if the feed loss is 0.2 dB and the antenna noise temperature, without considering feed loss, is 35K under clear-sky conditions.

Solution.

$$T_{A/f} = 10^{-0.02} \cdot 35 + 290\left(1 - 10^{-0.02}\right) = 46.5 \text{K}$$

List of Acronyms and Abbreviations

ACI   adjacent channel interference

ADC   analog-to-digital conversion

ADSL   asymmetric digital subscriber loop

AGC   automatic gain control

AM   amplitude modulation

AMSS   Aeronautical Mobile Satellite Services

ATM   asynchronous transfer mode

ATSC   Advanced Television Standards Committee (United States)

AWGN   additive white Gaussian noise

AZ°   azimuth (in degrees)

BAT   bouquet association table

B   bandwidth

BER   bit error ratio

BOL   beginning of life

BPSK   binary phase shift-keyed

BSS   broadcasting satellite system

BW°   half-power beamwidth (in degrees)

CA   conditional access

CATV   cable television

CCI   cochannel interference

Codec   encoder-decoder

C/I   carrier-to-interference ratio

C/N   carrier-to-noise ratio

COFDM   coded orthogonal frequency-division multiplex

CONUS   contiguous United States

DAB   digital audio broadcasting

DAVIC   Digital Audio Video Interactive Council

dB   decibel

dBi   decibel relative to an isotropic source

dBm   decibel milliwatt

DBS   direct broadcast by satellite

dBW   decibel watt

dc   direct current

DCT   discrete cosine transform

DEM   demodulator

DRO   dielectric resonant oscillator

DTH   direct-to-home

DVB   digital video broadcasting

DVB-CI   digital video broadcasting–common interface

DVB-S   digital video broadcasting–satellite

DVB-SI   digital video broadcasting–service information

DVB-T   digital video broadcasting–terrestrial

EBU   European Broadcasting Union

ECM   entitlement control message

EHF   extremely high frequency

EIRP   equivalent isotropic radiated power

EIT   event information table

EL°   elevation angle (in degrees)

ELG   European Launching Group

EMM   entitlement management message

END   equivalent noise degradation

EOL   end of life

EP European project

EPG electronic program guide

ETS European Telecommunication Standard

ETSI European Telecommunication Standards Institute

EUTELSAT European Telecommunication Satellite Organization

FDM frequency division multiplex

FDMA frequency division multiple access

FEC forward error correction

FET field effect transistor

FM frequency modulation

FSS fixed satellite services

GaAs gallium arsenide

G/T the figure of merit for a receiving system

GEO geostationary orbit

H horizontal (polarization)

HDTV high-definition television

HEMT high-electron mobility transistor

HPA high-power amplifier

HPS high-power systems

HVS human visual system

I   in-phase, or interference, interferer

IBO   input backoff

IF   intermediate frequency

IFA   intermediate frequency amplifier

IMUX   input multiplexer

INMARSAT   International Maritime Satellite Organization

INTELSAT   International Telecommunication Satellite Organization

INTERSPUTNIK   International Space Telecommunication Organization

IP   Internet protocol

IRD   integrated receiver and decoder

ISDB   integrated services digital broadcasting

ISO   International Standards Organization

ITU   International Telecommunication Union

ITU-D   ITU Development Sector

ITU-R   ITU Radiocommunications

ITU-T   ITU Telecommunication Standardization Sector

JPEG   Joint Photograph Expert Group

K   Kelvin (degree of absolute temperature)

K_p   proportionality's constant

k   Boltzmann's constant

LDTV   low-definition television

LEO   low Earth orbit

LHCP   left-hand circular polarization

LMSS   land mobile satellite service

LNA   low-noise amplifier

LNB   low-noise block (LNA plus downconverter)

LNBF   low-noise block and feed

MAC   multiplexed analog components

MMSS   maritime mobile satellite system

MOD   modulator

MoU   memorandum of understanding

MPEG   Motion Picture Expert Group

MPS   medium-power system

MSS   mobile satellite service

MUX   multiplexer

NIT   network information table

NOC   network operating center

NTSC   National Television System Committee

OBO   output backoff

OBP   onboard processing

OEM original equipment manufacturer

OI orbital interference

O-QPSK offset QPSK

OMUX output multiplexer

PAL phase alternating line

PAT program association table

PC personal computer

PCI personal computer interface

PID packet identifier

PFD power flux density

PLL phase-locked loop

PMT program map table

PRBS pseudorandom binary sequence

PSD power spectral density

PSI program service information

PSTN public switched telephone network

Q quadrature, quality factor of a tuning circuit

QEF quasi-error-free

QoS quality of service

QPSK quadrature phase shift keying

RARC   Regional Administrative Radio Conference

RF   radio frequency

RHCP   right-hand circular polarization

RS   Reed-Solomon code

RST   running status table

R_x receiver

SAW   surface acoustic wave

SDH   synchronous digital hierarchy

SDT   system descriptor table

SDTV   standard definition television

SECAM   Sequential Couleurs à Memoire

SFD   saturated flux density

SHF   super high frequency

SI   service information

SMATV   satellite master antenna television

SMS   subscriber management system

SONET   synchronous optical network

SSPA   solid-state power amplifier

ST   stuffing table

S_x baseband signal

TCP   transport control protocol

TDM   time division multiplex

TDMA   time division multiple access

TDT   time and data table

TI   terrestrial interference

TS   transport stream

TT&C   telemetry, tracking, and command

TV   television

TWTA   traveling wave tube amplifier

T$_x$ transmitter

UHF   ultrahigh frequency

V   vertical (polarization)

VSWR   voltage standing wave ratio

W   watt

WARC   World Administrative Radio Conference (also WRC)

XPI   cross-polar isolation

X$_{POL}$ cross-polar interference

About the Author

Jorge Matos Gómez graduated with a degree in electrical engineering and a major in telecommunications from the Universidad Central de Las Villas (UCLV) in 1970. He was an instructor in the Facultad de Ingenieria Electrica, UCLV's Department of Electronics and Telecommunications. At the same time he worked on his dissertation at UCLV—in collaboration with Professor Fritz Wiegmann at Technische Universität in Dresden, Germany—and earned a Ph.D. in 1982. He has taught many subjects related to the field of radiocommunication engineering, including radiocommunication system engineering, radioelectronics, stochastic process and probability, analog and digital communication systems, antennas and propagation, microwave engineering, and electromagnetic theory. He has also worked in masters and doctoral programs on the subject of satellite systems engineering (with applications in direct broadcasting). He has served as a professor in UCLV's Department of Electronics and Telecommunications since 1991. In addition, he has engaged in professional duties at several universities in Latin America, and in 1997 he worked in the master's program on space radiocommunications under the tutelage of Professor Gerard Maral at the Toulouse, France, site of École Nationale Supérieure des Télécommunications. He is currently working on a joint research project with Decom–Faculdade de Engenharia Elétrica e de Computção–Universidade Estadual de Campinas, Brazil, in the field of digital broadcasting by satellite and terrestrial means.

Professor Matos Gómez has published or presented more than 25 papers in scientific journals and at international events on telecommunications. His main professional areas of interest are related to coded modulation applications, digital receiver architectures, and radio propagation models used in the Ku band for direct-to-user satellite systems applications. Professor Matos Gómez has also been involved in curricular design activities in undergraduate and graduate programs in telecommunications.

Index